Managing Metrication in Business and Industry

American National Metric Council
1625 Massachusetts Avenue NW
Washington, D.C. 20036

Managing Metrication in Business and Industry

MARCEL DEKKER INC.
New York and Basel

The primary function of the American National Metric Council (ANMC) is to coordinate on a national scale voluntary metric activities in and among various segments of the economy. The need for planning and management of the voluntary implementation of metric is the cornerstone of this effort. This publication was written to assist companies and organizations in meeting the planning and management challenges of metric conversion—whether their decision is to adopt metric units or to continue with the customary measurement system. Planning and management are necessary to arrive at and implement either decision.

This book is the result of a carefully planned and organized ANMC Metric Management Workshop held in Chicago on October 6 and 7, 1975. The case-study approach to metrication management was used because there is no "right" approach to the opportunities and problems that must be faced when considering metric conversion. Each company must plan and manage according to its own particular set of circumstances.

This publication is not intended to be the final testament on metric conversion. *Managing Metrication in Business and Industry* is meant to be thought provoking and to illustrate the common questions and problems of metrication management. Again, the case-study approach is appropriate because no one company will embody all of the necessary topics in equal detail. Our fictitious company, Cirtem Corporation (Cirtem is metric spelled backwards), was intentionally structured to cover the full gamut of issues in metrication as an aid to the reader in approaching his specific situation.

The Metric Management Workshop was conducted in the following manner: background papers and case studies were prepared and read by the participants in advance of the workshop. Each of the eight topics was then discussed with the aim of achieving a consensus on what Cirtem's action in each area should be. *Managing Metrication in Business and Industry* is not a chronology of these discussions; it is a distillation of their highlights.

In using this book, it is recommended that the Introduction and then the Part III summary be read first. This will provide an overview and summary of the key issues and some possible approaches. Part III is a summary of the discussions by the workshop moderators, distinguished professionals who are involved in the day-to-day metric activities of organizations ranging from hardware manufacturers to consumer goods producers. The busy executive will therefore find the Part III summary especially valuable in understanding the basic issues of metrication, for the flavor of actual experience permeates each summary. The background papers presented in Part I introduce the reader to the major issues that must be addressed in the case studies, which are presented in Part II. Using this method, the reader can select the topics of greatest interest and pursue them to the appropriate level of detail.

Managing Metrication in Business and Industry was written for management, metric coordinators, engineers, financial personnel, computer operators, manufacturing and procurement experts, designers, and training, marketing, and public relations people who have a desire to keep pace with private sector metrication activity. It is also useful for federal, state, and local government officials, because planning and managing metrication is equally important to the public sector.

ANMC expresses its appreciation to the workshop moderators for their time and effort and to their organizations which supported them. Special thanks to each of the participants of ANMC's first Metric Management Workshop for their contributions and permission to list them in Part III. The Council would also like to acknowledge the work done by the editors of the book, Len Boselovic and Lou Perica.

INTRODUCTION

MALCOLM E. O'HAGAN

President
American National Metric Council
Washington, D.C.

On December 23, 1975, President Ford signed the Metric Conversion Act of 1975 into law. The Act calls for a "national policy of coordinating the increasing (voluntary) use of the metric system in the United States." The law does not prescribe a time period for conversion; rather, it states that each sector should convert when it becomes economically feasible to do so.

For some, however, metric legislation is coming after the fact. The world-wide trend to SI metric has made U.S. conversion inevitable. In both the public and private sectors, planning for metric conversion has been underway for some time.

Many federal agencies have already begun to prepare for the changeover. The Treasury Department has issued regulations requiring metric-sized wine bottles by January 1, 1979. This directive will reduce the number of domestic wine bottle sizes from 16 to 7, and the number of imported bottle sizes from 27 to 7. The Environmental Protection Agency requires all new standards and regulations to be written in metric units. The Office of Education has awarded $2 million in metric education grants. Finally, a June 1975 Department of Defense (DOD) directive recognizes U.S. metric conversion as inevitable, and states that DOD must go metric when it becomes economically practical.

In the private sector, major U.S. corporations have included a coordinated, cost-effective conversion among their corporate goals. Dozens of the nation's largest corporations are planning for the transition, and in some cases implementation is already underway.

The automobile, off-road vehicle, computer, and steel industries are in the vanguard of the change. But serious metric planning has begun in most industries, with the large multinational companies setting the pace. The action of corporate giants is having a precipitous impact on their thousands of suppliers, and a catalytic effect on other corporations.

Why are so many U.S. corporations already planning for metric conversion? Reasons vary from one corporation to another depending on the nature of their business. Willard F. Rockwell, chairman of the board of Rockwell International, stated the rationale very well in an address to ANMC's 1st Annual Conference in March 1975. Calling metrication a "once-in-a-lifetime opportunity for American industry," Rockwell said use of the metric system "can expand U.S. markets, improve productivity, reduce costly inventories, and cut production costs." He further noted that experience at Rockwell International has shown that there are potential advantages to metrication: simplification of international activities; expansion of foreign markets; expansion of supplier choices; and flexibility in fabrication location.

Rockwell also stated that "We can see considerable future advantages in the formulation and adoption of well-thought-out standards which can reduce the number of parts, tools, and gages required. These will produce more efficient designs and practices. They'll offer an excellent opportunity to "clean house" by eliminating many near-duplicate sizes. And it is in that regard I believe overall national conversion to the metric system could be a once-in-a-lifetime opportunity for American industry."

Two facts about metrication, then, confront U.S. industry: (1) conversion is inevitable; and (2) it offers opportunities as well as problems. Because of the first, it is only good business sense to make the most of the situation, to resolve the problems and realize the opportunities through careful planning and management.

Realizing that metrication is inevitable and recognizing the need for management, private industry has established and is supporting its own metric coordinating body: the American National Metric Council. As the manager of metric conversion, the Metric Council is helping the private sector undertake the changeover wherever and whenever it becomes desirable. ANMC is nonadvocate: it does not attempt to convince anyone that metric conversion is necessary or that it should be resisted. However, ANMC does advocate a well-planned, well-managed, and nationally coordinated approach to metrication on the part of those deciding to make the change.

ANMC's *Managing Metrication in Business and Industry* presents an overall approach to a well-planned company metric conversion program. It offers guidelines in eight major functional areas affected by metrication, as seen through the eyes of corporate officials who are presently managing the change. The case-study approach to the subject illustrates that metrication presents different planning and management challenges to each corporation, that what is right for one company may not necessarily be right for another.

More importantly, the book is based on the fact that metrication is something that U.S. industry must come to grips with, that it cannot be met with a "wait and see" attitude. Metric conversion requires a well-conceived, carefully developed corporate plan. *Managing Metrication in Business and Industry* offers the first step in that planning and management effort.

CONTENTS

Part III Case Study Issues Identified

Part IV Metrication Cost Management: What Others Are Doing

Part I

FACING THE ISSUES OF METRICATION

Each chapter in this section provides the information necessary to address the issues raised in the case studies presented in Part II. These chapters are not exhaustive or comprehensive presentations on the subject matter. They are intended to provide the background and guidance necessary to arrive at approaches and solutions for the problem facing the hypothetical Cirtem Corporation: whether and how to go metric. This information can be adapted by the reader to fit the specific needs of his organization.

Chapter 1

COMPANY METRICATION STRATEGY AND PLANNING GUIDELINES

LOWELL W. FOSTER

Corporate Standardization
Honeywell, Inc.
Minneapolis, Minnesota

The mere mention of the word metrication is likely to initiate a heated discussion. If a company is seriously contemplating the metric issue, however, it soon becomes evident that discussion must give way to astute planning, including strategies and guidelines in approaching this important subject.

Early Planning

Normal good business management principles are, of course, assumed. Yet, it seems that the fear of such an apparently mountainous task can tend to dull the normal business sense. Metrication appears as a "monster" of sorts, and thus can tip the balance toward confusion and misunderstanding in viewing the potential of this once-in-a-lifetime experience. Normal business instincts look for the payoff; metrication, at least in the short term, is jealous of revealing that real payoff. However, metrication offers significant long-term benefits, and they can only be attained through exhaustive investigation and careful planning.

Companies normally operate under two forms of management: "horizontal," which is the foundation of basic resources, practices, and policies;

and "vertical," the product-oriented needs to perform specific tasks in getting its profit cycle successfully completed with parts out the door. Thus, the product manager whose instincts tell him that metric needs should be developed only when those needs arise in building the product, may be in for a surprise. This will not be the case if some concern has been given by others for developing the "horizontal" resources and setting them in place to do the job. Therefore, slightly more imaginative management and planning principles may be crucial to a good metric program. Planning ahead and being prepared, not totally of course, but with the essentials in place, can be extremely important.

The principles of leadership or "followship" are also weighed at this point. It is not really possible to make a clear choice. However, a clearly mandated program which addresses the real future needs of the company, just as it addresses research, will make the company neither a leader nor a follower but a dynamic participant in the wave of progress underway. To falter too long (be a follower) may bring a serious blow when it is discovered that metric progress must be grown—it cannot be bought.

A metric plan, then, seems to naturally evolve out of any serious metric anticipation. It is also natural to assume that each company will require some uniqueness to its planning even though common threads of uniformity do exist. For example, Honeywell's planning and guidelines evolved out of its needs as a multinational company and its concern for its industries and U.S. progress toward metric evolution.

Honeywell's Decision

Following are references to some of the thoughts, motivations, and considerations which led to Honeywell's decision to embark upon its formal metric program beginning July 1, 1972. Influential elements in that decision were (1) the company's past history of building some metric products for foreign sales; (2) U.S. metric progress, and (3) need for improved standardization. The Honeywell Plan, which is discussed below, will summarize some of the key factors which moved the company ahead on this metric road.

The company metrication plan covers a general 10-year period. The essence of the plan is to provide a coordinated company position of awareness, preparedness, and rationale on metrication. This minimizes costs and confusion and maximizes efficient management in the metric transition period. The plan is not intended to cause precipitous actions, but rather to prevent, as much as possible, such actions.

Considering the countless variables involved, establishment of an extremely rigid plan would be foolish. Therefore, the plan should be constructed with built-in flexibility. This permits necessary adjustments as events unfold. The plan provides a beginning, a point of departure, from which the future can be visualized and the present can be guided.

Basic to effective planning is setting priorities during the entire transitional period. Product or internal needs are first, followed by the need to contribute to the U.S. and worldwide industrial scene.

The Honeywell Metrication Plan covers 10 years. It was activated on July 1, 1972. Its basic objective is to achieve "full metric capability"; in essence, to evolve into predominant metric use in that time. Recognizing the advisability for planning and implementing the metric transition in stages, the 10-year plan is divided into the following phases:

Phase A—Metrication Alert (2 years): A period in which certain metric considerations in design, production, purchasing, and marketing are made which build a state of preparedness.

Phase B—Metrication Adaptation (3 years): A period in which specific adaptations are made in keeping with increased company needs, established national policy, and changing industry conditions. In some situations, these adaptations will approach 50% metric applicability.

Phase C—Metrication Applicability (5 years): A period in which metrication will be vigorously implemented. Phases A and B will be fully considered as Honeywell becomes a totally metric company.

Further refinement of requirements and priorities may be necessary within divisions during the metric transition. Any considerations made by divisions or groups should dovetail as much as possible with the overall plan.

Full metric capability is defined to mean compliance with *all* of the following conditions:

1. Complete understanding of the SI system
2. Maximum application under the constraints of need and economic feasibility.
3. A state of metric accomplishment abreast of industry and overall U.S. progress expected in 1982

Coordinating the Plan

Where a division or group determines the time is right to proceed into metric, it is guided by the overall plan and sets its own priorities for the various

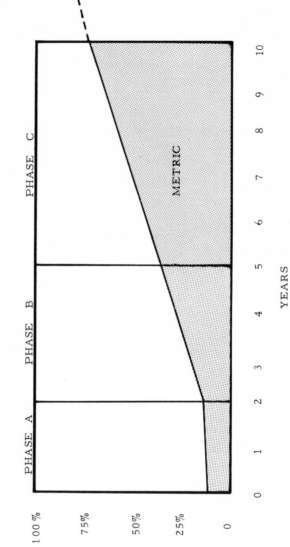

GOAL: REVERSE ROLE OF INCH/METRIC SYSTEMS
IN U.S. OPERATIONS

FIG. 1. Corporate metrication plan.

6

operating parts of the division. These are, for example, engineering, design, manufacturing, inspection, procurement, marketing, finance, standards, training, sales, research, tax, personnel, etc. If necessary or desirable, a metrication coordinator and a metrication committee are established to formulate specific division guidelines. Coordination with the corporate plan via the "metrication nerve center" at the corporate standardization headquarters is pursued at the discretion of the divisional management and metrication team. Technical or managerial assistance, coordination, training capability, etc. are available at corporate headquarters and at key locations around the corporation.

Figure 1 illustrates the composite corporate Honeywell Metrication Plan. The goal is summarized as "Reverse Role of Inch/Metric Systems in U.S. Operations." Beginning with nominal application of metric in 1972, the plan progresses to "predominantly" metric application in design and build in approximately 1982.

The focus of attention during the transition will be U.S.-based facilities. However, the planning is also assumed to coordinate with Canadian operations and with other international divisions, such as UK and Australia. Specific consideration is also given to divisions in established metric countries where transition to the new SI units is also required. Obviously, varying conditions exist, and it is the responsibility of the concerned operating units to use their own discretion in best interfacing with others.

Responsibility for metric implementation and coordination with other units of the company lies with the division or groups concerned. Refined division or group planning is anticipated as necessary, and will be based upon each unit's economic feasibility and guidelines.

Authority for implementation of the metrication plan lies within the corporate policy and practices established in conjunction with the plan and as authorized by top management of the company.

Figure 2 illustrates how a division or unit of the company might set its own planning within the corporate plan and Phases A, B, C. The single cross-hatched area represents "begin metric transition toward 50% capability." This is the metric indoctrination period. The double crosshatched area is the period "extending over 50% metric capability," and the solid area is the time for "approaching full metric capability." Note that these periods straddle the 10-year plan (phases A, B, and C) but accent the detailed priorities.

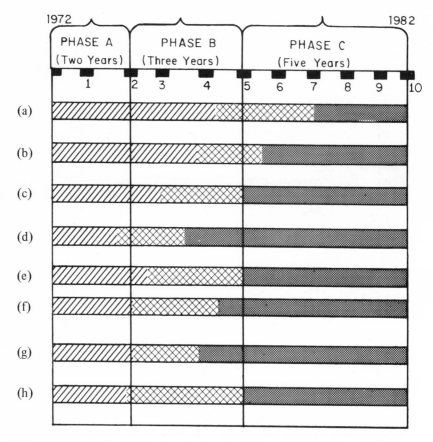

FIG. 2. Metrication plan chart. (a) engineering design and documentation;
(b) manufacturing; (c) procurement and materials; (d) training; (e) standards;
(f) marketing and sales; (g) research and development; and (h) legal and tax.

Engineering Design

Engineering design and documentation is one of the most important aspects
of metrication within a company. It is at the design stage that selection of
sizes, characteristics, etc. of the product are established. Designing to metric
modules can be readily accomplished if metric standards and other needed
data are available. Development of key standards and material data to
incorporate both inch and metric equivalents, and to indicate where standard
tolerances overlap can facilitate metric design. Development of engineering

and application standards of this kind within the company can be given priority consideration. Further, raw materials and the manufacturing capability to work those raw materials into metric products are necessary to assure the cycle from design to finished product. Designing in truly metric terms becomes practical only when the necessary supply of metric tools are available. Metric tools include the standards, data, and, of course, the physical tools and human resources.

Metric design can be introduced in some products using the metric modular concept. In this instance, the outside dimensions and mounting features are metric to fit or interface with another metric device. The internal design and components, however, may be inch-based to no detriment of the functioning of the device.

Manufacturing

In manufacturing operations (cutting, forming, etc.), the actual production molding or casting of a metric part presents little problem. Adjustment of depth of cut or selection of standard tools can, in most cases, easily produce metric sizes.

The human aspects of complying with the engineering requirements, producing the tools, and setting the machine dials properly, however, present other considerations. Conversions in one form or another (dual dimensions, conversion charts, calculations, etc.) are necessary. This requires time and introduces error possibilities.

Conversions of machines and equipment and procurement of new equipment should be carefully studied and planned. For example, metric machine and equipment buying may be accomplished through regular replacement programs with the possibility of little, if any, added cost. Dual (metric and inch) graduated and capable machines or conversion kits are also available as other possibilities.

The sometimes held impression that metric conversion requires wholesale replacement of machines and equipment is entirely false. One of Honeywell's UK factories, for example, accomplished all the necessary conversion of its existing manufacturing equipment in a relatively short period by doing the work with its own shop people and with little added cost, loss of time, or confusion. The only replacements or new equipment came as normally required in the capital equipment replacement program.

Micrometers, calipers, verniers, and similar items should be procured as necessary. Conversion of small tools to metric is impractical. Electrical, pneumatic, pressure, etc. calibration in production, inspection, or evaluation equipment can, in most cases, be recalibrated or reworked to provide any necessary dual capability.

Crucial to successful metrication transition is the matter of standards and their availability to guide metric product design and production. Top priority should be given to standards development and availability as a basis for metric implementation. Of primary concern should be fastener standards, material thicknesses, etc. Adaptation to metric standards by overlapping tolerance ranges (inch/metric) may open up early possibilities. Concerted efforts in these and other areas are now underway within Honeywell, and in ANSI and ISO circles.

Metrication is really only a part of a much larger concern: international standards. United States involvement in the development of international standards is crucial to future United States trade. Standards developed without United States input can result in trade and communication barriers which seriously impair the United States influence and advantage.

Procurement

Procurement plays a very important role in the metric transition program. It must keep the normal pipelines of supply open, provide maximum economies and availability, and also provide the expertise in searching out new metric resources. Procurement should work closely with management, planning, marketing, and engineering the transition, and with standardization in planning and forecasting the material needs of the future. Procurement should also screen orders for capital equipment as to whether a metric capable unit is desirable. Dual inventory of raw materials, parts, small tools, and expendable supplies in both inch and metric base is an item of serious concern and requires careful management and planning. Procurement lends much assistance in helping to monitor orders and by applying effective and cost-conscious management to supply pipelines.

Training

The success of a metrication program rests, finally, on the people involved and how successfully they respond to the technical, managerial, and psychological adjustments conversion requires. Due to the human emotion involved, metrication provides a unique relationship between the technical and psychological aspects. Since there is human resistance to change, training on the subject of metrication takes on importance as a key item.

Metric training should have a low-key emphasis, and employees should be taught only what they need to know. Only persons who understand the company, its products, and the needs of the people involved should act as trainers. Finally, the training effort should be coordinated throughout the company. Training aids and programs should be shared and capable company instructors should be pooled to make the most effective use of training resources and to minimize duplication of effort.

Early Company Metric Interest

The foregoing are some of the factors that resulted in the "new generation" metric interest and in Honeywell's metric conversion plan. (Further discussion of these forces is presented later in this chapter.) Earlier company history on the subject was also instrumental in causing company progress on metrication. It is, in fact, possible to state that the company's astuteness on this subject is not accidental. It is the product of a long history of metric exposure and the monitoring of the worldwide metrication movement.

As a result of its role as a major supplier, contractor, and manufacturer of goods and material during and following World War II, Honeywell developed a reputation as a rising multinational company. Early (1940s and 1950s) affiliation with foreign divisions and businesses led the company to build some metric parts in the U.S. and some totally metric devices in its European plants. Thus, the sense of a metric product potential was as natural as Honeywell's aspirations to grow in the worldwide marketplace.

Interface manufacturing was typical of Honeywell's early U.S. metric manufacturing activities (for example, a valve control designed and manufactured to inch design, but fitted with metric-size thread mountings and built-in adjustments for variable pressures to suit performance and testing criteria in foreign countries). This was Honeywell's baptism of metrication.

Manufacturing of metric products was, of course, routine in metric countries such as West Germany, Holland, and France. In the early days, this metric experience was shielded from U.S. division interest. However, with the increasing transfer of work, personnel, product marketing, etc., this shield has gradually dropped and a wider exposure now adds to Honeywell's metric momentum. For example, thermostats, once designed and manufactured for metric markets, are now being considered for marketing in new metric nations (including the U.S.).

Honeywell Metric Studies: 1962–1971

Also of particular value and interest are some company metric studies undertaken in the 1960s. Each of these events can be counted as significant to Honeywell's positive posture and ready acceptance of metrication.

In 1962, the first action concerned with the possible formal metric conversion within the company (and the U.S.) was undertaken. At the request of executive management, a metrication committee of six individuals was charged with studying the possible implications of metric conversion; that is, if it were to occur in the "sometime" future. This committee was composed of corporate staff personnel and residential division (automatic controls division). After three months of spasmodic efforts, a report of a very general nature was submitted to upper management. The report removed some of the fears about conversion, clarified some of the issues, set somewhat of an optimistic tone, but left the impression that more in-depth study was required.

In early 1963, the author was assigned the task of conducting a more in-depth metric study, but within a limited time and resource constraint. It was to be done "solo" and within approximately 40 working hours. The job was ultimately accomplished within the foregoing dimensions, but with the voluntary involvement of numerous resource persons at various positions within the company. A brief report of impressions, methods, objectives, and results of the 1963 study is of interest.

After initial consideration, it was decided that the report would:

1. Be cost-and result-oriented
2. Be based on a 10-year conversion period
3. Assess the major functions impacted, especially the affects on design and production
4. Establish a company position on the subject
5. Be based on available or speculated "facts" and written in terms of rations or percentage figures so that a sliding scale criteria could serve in future assessment or up-dating

The first step was to gather data, to study the impact and possibilities of metric manufacture (could it be done with existing equipment?), look at the raw material and supply problems (e.g., dual inventories and dual documentation), talk to knowledgeable and concerned people to ascertain the psychological effects, etc. Needless to say, this was a pioneering effort undertaken with fear and trepidation, a challenge to be sure. However, the results of this study were both surprising and encouraging.

Surprising, because of some 50 people contacted and burdened somewhat by the uniqueness and weight of the questions asked, *no one* seemed opposed to the idea of metric conversion. Remembering that this was 1963, when metrication was far from the buzz word it is today, is even more surprising today than it was then. Granted, some had an "it can't happen here" attitude. Nevertheless, only interest, optimism, and curiosity seemed evident in their reactions. However, there was a great concern on the part of some as to when this great event was to be expected. Surely, they said, you do not encourage the change now. The assurance was given that it was only a study, but the participants were told that speculation as to the future was serious.

The cooperation received from everyone involved in the study as resource persons was excellent. Facts and figures were available on such things as metric conversion of machines and equipment. Other areas, such as the costs of dual documentation in the transition period and the training of people, required an "educated guess."

In the categories of design and documentation, modification of equipment, training, etc. (shown in Fig. 3), individual studies of cause and effect, time, and costs were determined on the basis of a 10-year conversion plan. It was said then, and it is repeated loudly now, that the cost results were (and are) somewhat crude. Any price tag on metrication made before the fact must be crude. However, a later (1971) study, which is discussed below, corroborated the original estimates. The early study did isolate the areas of concern, set some priorities, and gave a "ball park" evaluation.

Figure 3 reveals the cost impact results of the 1963 study. Each of the areas used as key concerns is listed with a percentage and dollar value assigned. Totaling 3.63% or $3.63 million, cost over the 10-year period represents a "then assumed" optimistic figure. Only one or two other companies to our knowledge had made any similar studies at that time (1963), and their figures were in the 15% to 20% range, with a very rubbery base (obviously).

Note that the tabulated results of the 1963 study shown in Fig. 3 are based upon a hypothetical figure of $100 million gross sales. This permitted the sponsoring division to plug in its true sales figures at any time in the future (withing approximately a 5-year period) and derive a metric cost figure

Company "X"[a]

	% of Gross Sales
Modification or replacement of machines and equipment	($1 380 000)	1.38
Maintaining dual supply tools and facilities	(750 000)	.75
Modification or replacement of gages and testing equipment	($580 000)	.58
Redocumentation	($500 000)	.5
Education and Training	($20 000)	.42
Totals	$3 630 000	3.63%

[a]This study is based on $100 million gross annual sales year. Total cost for 10-year transition to metric use on basis of any one point in time, e.g. start 10-year transition program in 1972 and end 1982 (as based on 1972 $ value), est cost = $3,630,000.

FIG. 3. Results of 1963 Honeywell metrication cost study.

for the speculated 10-year conversion period. The 10-year period cycle was envisioned as shown in Fig. 4.

The report satisfied management needs for setting some direction, evaluating the impact, and determining company posture on metric conversion.

The years from 1963 to 1968 passed with monitoring of metric progress, both within the company (then drawing in more U.S. divisions in its concerns), in the U.S., and worldwide. Participation by the author and others in early national metric activities, such as the ANSI Metric Advisory Committee, supplemented Honeywell's activities.

Progress and interest from 1968 through 1971 were accelerated by participation, as a company, in the NBS Metric Study and increased involvement with multinational goals and worldwide markets. For Honeywell, metrication was no longer a question of "if"; it was a matter of "how" and "when."

In 1971, a detailed status report and policy and plan guidelines were developed for submittal to executive management. Also, as corroborating support, a restudy was made of the cost effect of metrication. The 1971 results, using the same base ingredients as the 1963 report, are shown in Fig. 5. Note that with the advantage of some experience, "hindsight," and

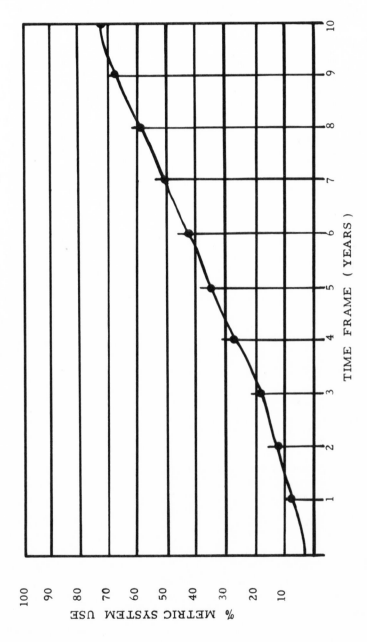

FIG. 4. Honeywell's 10 year conversion cycle.

15

Cost Estimate Table Division "X"[a]

	Costs 10 Year Total	% of Annual Gross Sales
Design and Documentation	($250 000)	.25
Modification or replacement of machines and equipment	($150 000)	.15
Maintaining dual supply of tools, parts, and materials	($250 000)	.25
Modification or replacement of testing equipment and gages	($210 000)	.21
Education and training	($160 000)	.16
Miscellaneous (marketing, sales, legal, tax)	($100 000)	.10
Total	$1 120 000	1.12%

[a]Based on $100 million gross annual sales over 10 year period.

FIG. 5. Results of 1971 Honeywell metrication cost study.

company divisional experiences in U.K., the margins were narrowed to a cost effect of 1.12% of gross sales for one year, stretched over a speculated 10-year transition.

These latter figures (Fig. 5) are admittedly crude; there are many implications to conversion and their cost is not predictable to any degree of accuracy. However, the figures served a purpose in helping to identify concerns and set priorities.

The popular thought today is to avoid attempting to isolate costs; it is thought to be an exercise in futility. We agree with this thought in principle. However, as good managers, some cost data seem essential to create the assurance that we are in control. In actuality, the more germane question, not appropriately approached as yet, is: what are the cost advantages of metrication? We are convinced that they are impressive to the point of overshadowing costs.

Honeywell's optimism toward metrication is a product of a general feeling that conversion can be accomplished within tolerable cost limits. The company believes metrication offers significant advantages in product expansion, new marketplace horizons, and better management of a worldwide corporation as a result of a worldwide language of measurement.

Company Planning 1971-1972

Following are key thoughts and statements of position which contributed to formulating the original Honeywell plan. Abbreviated summaries are used to convey the essence of approach to the subjects during 1971-1972 planning.

Company Metrication Position

It shall be the company position that a gradual and careful transition toward full metric capability be pursued in its design, manufacture, and marketing of products and be accomplished within 10 years. This period is to begin July 1, 1972.

Responsibility

This transition period will be a planned and coordinated one, where a "Metrication Plan" provides the guidelines and basic considerations to be made. The responsibility for implementation of the plan applicability or extent lies with the division or operating units concerned. Expanded or refined division or unit planning is expected as necessary.

Applicability

Applicability of the metric transition period and planning is directed primarily to U.S.-based facilities. However, the transition and planning will be assumed to also integrate with those international divisions, such as UK and Canada, now merging toward metric, in keeping with national policy, and all others which are now metric.

Authority

Authority for implementation and actions of the Metrication Plan lies within the broad base of the established Corporate Policy statement and Corporate Procedure.

The technical authority as the basis for company use of the metric system is found in the document, International Standards Organization (ISO) 1000, "International System of Units" (latest issue).

Policy

Honeywell will accept and support any subsequent national policy on metrication insofar as it is economic and practical to do so and in keeping with industry trends toward increased use of metric units of measure in the United States.

Worldwide activities of the Corporation require use of a universally accepted system of units of measurement and the Corporation will therefore adopt, from the International Organization for Standardization (ISO) SI "International System of Units," those units that are acceptable for common use as the base for interchange of technical data with divisions in metric countries or as appropriate to a changing U.S. industrial environment.

Company implementation of this policy in the United States shall include an appropriate plan, and shall occur in keeping with national policy and trends and as metric raw material, supplies, and basic engineering commodities become available. Implementation of this policy in divisions outside of the United States shall be in keeping with Corporate goals, and trends in their own areas as well as in the United States.

Coordination of this implementation shall be the responsibility of the Corporate Standardization Services Department in cooperation with the Interdivisional Standards Council—Metrication Task Force.

Cost Planning

Implementation Costs

Costs for implementation of the metric transition are generally assumed to be absorbed by the concerned division or unit through its regular operating budget. A general policy that "costs lie where they fall" is assumed.

Approximate Nature of Estimating
Metric Costs

Any attempt at calculation of costs relative to a long-range metric transition must be considered a broad and very crude approximation. Accuracy would depend upon a knowledge of those conditions which will prevail in the months and years ahead. However, by assuming a steadily continuing trend of industry conditions, such as increasing use of metric materials and gradually increasing metric manufacturing capability, some crude cost estmates can be attempted. This is the basis upon which the cost evaluation is made. Any drastic acceleration of metric priorities, such as caused by national policy, abruptly changing company needs, industry trends, etc., would significantly alter predictions.

Metric Costs Not Necessarily Added Costs

Costs in the context of the metric transition should, however, be visualized as a singling out of those costs attributable to the effort in becoming gradually metricated, and not necessarily as added costs. The gradual approach promoted in the Plan indicates numerous instances where the expenditure is a part of near normal operation to design and build. The cost would be expended to do the job in any system of measurement.

Savings Not Considered

No attempt is made in the Plan or in Cost Planning to analyze the possibilities of savings or advantages resulting from increased use of the metric system in our products. Certainly standardization to bring us in keeping with the predominant world measuring system, use of a common drawing language in units of measure, elimination of some redesign and redraw, closer adaptation to worldwide metric material and tool standards, etc. can be considered forms of advantage. These matters are, of course, woven into our planning and will have an effect on progress and the steps to be taken as the future unfolds.

Cost Estimate Table

The Cost Estimate (Fig. 5) addresses a hypothetical United States division of 100 million dollars gross sales per year and selects only major categories of concern. The dollar values are based upon the total estimated costs for each category over a 10-year period, since the gradual transition of the Plan covers 10 years. The percentage values are that part of annual gross sales (for one year). In this manner, costs for a single division or group of divisions for the 10-year period may be determined on the basis of annual sales dollars for that unit.

Plan

The Metrication Plan covers a 10-year period beginning July 1, 1972. The transition period has, technically, already begun, extending back some years through past and current metric requirements and involvement. The transition will continue with greater emphasis commensurate with growing needs and as aided and coordinated by the Plan.

In the absence of official United States national policy on this matter and in view of the variables concerned, establishment of a rigid plan at this point in time would be foolishly naive and irresponsible. Therefore, the Plan is assumed to be necessarily based on flexibility, permitting adjustments as future events unfold. The Plan provides a beginning from which the future can be built.

Phases A, B, and C

In keeping with a flexibility emphasis, the Metrication Plan is designed to apply in three general phases:

Phase A: Metrication Alert (Two Years)
Phase B: Metrication Adaptation (Three Years)
Phase C: Metrication Applicability (Five Years)

Plan Basis

The Metrication Plan is to cover a 10-year period based on an assessment of numerous considerations including company product needs, market potential, and corporate growth. The Plan is further designed to dovetail with an evaluation of currently rising national metric trends, evolving government interest, legislative considerations and actions, etc. Experiences within domestic divisions, exposure through the International Division efforts over the years, and most recently, the real experiences of change in the United Kingdom divisions of the company have also contributed to corporate planning.

Since planning is being done in the absence of an official United States national metrication policy and given numerous unknown future situations, speculation is necessary. However, Honeywell can proceed with reasonably high confidence that corporate planning will be in step with developments and will avoid any need for crash programs. We will, at the very least, be in a viable state of preparedness.

Corporate metrication planning is precipitated by two fundamental objectives:

1. Capability must be maintained and increased as necessary to meet ascending metric product and communication needs relative to worldwide business.
2. Honeywell must prepare for, and react to, national trends, legislation, policies, or actions which are occurring in the upswing of metric interest and use in the United States.

Expansion of company marketplaces around the world and the need to interface with multinational technologies requires a capability in both the inch/pound and the metric systems. To communicate and quantify necessary data on drawings, engineering and scientific documentation, standards, etc. to most of the world outside the United States and Canada, the metric language must be used. It is said that 95% of the world's population uses the metric language, while about 65% of the world's production is in the metric system.

Full Metric Capability

Full metric capability is defined, in the context of Honeywell planning, to mean complete coherence with the metric (SI) system as applied under the limits of economic feasibility, time, place, and situation, and not "metrication for metrication's sake." Full metric capability is likewise assumed to mean a state of metric accomplishment abreast of the national United States state of the art expected in 10 years. It does not mean total metrication to the exclusion of inch/pound base units, which are expected to remain in limited use indefinitely.

The full extent of Honeywell capability will become better focused as the events of time, national policy, and experience take place and permit adjustments. It is anticipated that corporate metrication planning, as guided by what appears to be an irreversible upward metric trend in the United States, will keep step with that trend. Metric capability in the interim is assumed to represent adequate progress within the guidelines of our plan commensurate with needs and equivalent to the general state of the art at any point in time.

Coordination

Coordination of the metrication program will be provided by the Corporate Standardization functions with the cooperation and input of the Inter-divisional Metrication Task Force.

These groups will provide a nerve center for assistance, information, plan adjustments, gathering of pertinent data, and monitoring of progress as necessary. The Corporate Standardization function will be the responsible location for the metrication nerve center. A semiannual review and adjustment of the Plan as necessary is considered essential to implementation and viability of our planning.

It is suggested that each division or unit involved in the implementation of the metrication plan coordinate its activities with others (within and outside the division) through its Corporate Standardization Committee representative or a specially assigned metrication administrator.

National Action

In 1968, a new dimension was added to the metric question in the United States and thus to company considerations on the matter. Action was initiated by the United States government under Public Law H.R. #90-472 to determine the advantages and disadvantages of increased use of the metric system in the United States. The National Bureau of Standards, under the Secretary of Commerce, conducted a three-year study (in which our company participated) to carry out the requirements of this congressional action. The results were published in July 1971 in NBS Report 345, "A Metric America— A Decision Whose Time Has Come," which also constituted Secretary of Commerce Maurice Stans' report to Congress on August 8, 1971.

The report was pro-metric and recommended that the United States embark on a 10-year coordinated program to increase its metric use toward making America a metric country. This is generally assumed to mean that the metric system (SI) will become the primary and officially preferred measurement system with the inch/pound system being acceptable, but in gradually declining use.

It is anticipated that eventual United States action will lead to the establishment of a top coordinating group (e.g., a National Metric Board along with subcommittees for various sectors of our society: law, industry, education, etc.). These groups would probably determine the overall planning, priorities, etc., and then coordinate the total effort. It is very doubtful that these groups would have strong mandating powers; coordination will probably be their major responsibility. Honeywell planning must anticipate the possibilities of government action and the manner in which its progress and program should be adjusted.

Attitude

Important to metric transition progress is a strong attempt to retain and emphasize a "business-as-usual" atmosphere. Metric planning should be devoid of emotion as much as possible and should be conducted with a distinctly low-key emphasis. Undue attention drawn to the metrication effort can create problems where none should exist and tip the balance needed in striving for a progressive yet stable situation.

One area in which the above emphasis should be seriously considered is metric training. A massive program in which many people are subjected to broad training would be extremely costly and detrimental to progress. Training should be handled at a gradual pace, accommodating only those needing the information during the transition; and then, most importantly, imparting only the extent of training needed by the individual to do his job. For most employees, it is not necessary to learn the entire SI system. Indoctrination covering the entire system in general terms is, however, excellent and should be used as a training emphasis, in earlier programs particularly.

Training people on a gradual and minimum basis can be considered as one of the "secrets" of a successful metrication program. This has evolved as an emphasized recommendation by the Honeywell UK metrication efforts and is based on their experiences. Their transition toward full metric capability has progressed smoothly with minimum problems using a very low-key training approach. Our own initial efforts in training of an indoctrination nature seem to have proven this theory. People seem to adapt and react cooperatively if not rushed into something new and strange. Build upon that which is already known and adaptation will follow quite naturally.

Metrication Status Report

A status report on metrication involvement within the company in the recent past was prepared and submitted as an important resource to support the Plan.

Conclusion

The metrication effort at Honeywell is a total and committed effort. We will accomplish our goals successfully through careful planning, coordination, and good management of all our resources. The goals and aspirations of our metric program are well worth striving for; any penalties, problems, or sacrifices do not appear to be intolerable. The final successes will come through our ability to connect the end objectives with the surest course of action. Course corrections are an important part of our planning. Therefore, with the ingredients enumerated and a clear vision ahead, Honeywell will achieve full metric capability and meet the needs of the future.

Chapter 2

METRICATION COMMITTEE FUNCTIONS

LARRY J. OUWERKERK

Hydraulics Division
Borg-Warner Corporation
Wooster, Ohio

Board of Directors Responsibilities

When the decision to go metric has been reached, one of two things will occur. Depending on the size of the company, the Board will either begin to piece together a plan for conversion, or it will appoint a Metric Task Force to begin the planning. However, no matter which of these approaches is adopted, the planning body must adopt an overall approach to metrication.

The decision to convert must have the full support of management. In addition, company directors should take the necessary steps to inform all personnel of metric planning. But the announcement should be low-keyed so that an overreaction to the plan does not take place. A majority of the people will know very little about the metric system, so it is important that the announcement does not make them fearful of the change.

Management will have to select an executive to guide the planning and implementation. This Metric Coordinator should be well versed in the workings of the firm and the products marketed. He should have the respect of managers who will be involved in the conversion, and should be able to maintain a working relationship between the Board of Directors and the Metric Committee. For some firms the job of directing the metric effort will be part-time, but in others it will require a full-time administrator.

The Board of Directors should approve metrication plans at various stages of development. This not only enforces the idea of management approval, but it will keep them informed of the status of the program.

Steering Committee Responsibilities

Corporations have a myriad of sizes and shapes. Each has its own personality. A large decentralized corporation may have several steering committees, whereas a more centralized firm may only have one. It is obvious that a smaller firm may not need a steering committee at all, in which case the metric coordinator would shoulder total responsibility.

In an organization of several divisions, the organization structure is shown in Fig. 6.

Regardless of the size of the company, every metric committee will have many general responsibilities. However, the large decentralized firm requires a more extensive metric organization, including a corporate metric coordinator. This official should be responsible for coordinating the metrication effort and should serve as a liaison between the directors and the divisional metric coordinators. Along with other specialists from corporate headquarters, the corporate metric coordinator should form a metric steering committee to (1) develop a working knowledge of metrication in all areas of company operations; (2) develop the transition plans for corporate headquarters; and (3) establish criteria for conversion in each division.

In addition to focusing on the details of conversion, the metric committee should consider the many opportunities that metrication offers. For example, the product lines involved should be examined to distinguish the money-makers from the losers. Eliminating the latter could mean longer production runs of fewer sizes, which translates into substantial savings. Metrication also offers the opportunity to redesign products. With technology expanding at an ever-increasing rate, now might be the right time to incorporate new technology in product design.

Legal aspects of metrication will also have to be investigated. In many instances, close cooperation must be maintained between vendors and the corporation, and customers and the corporation. Meetings could be considered violations of the antitrust laws. Should communication and cooperation be required between competitors, some organizations may take the opportunity to charge product allocation or price-fixing. Although there may be government legislation to provide protection to individual corporations, the ramifications should be made known and considered.

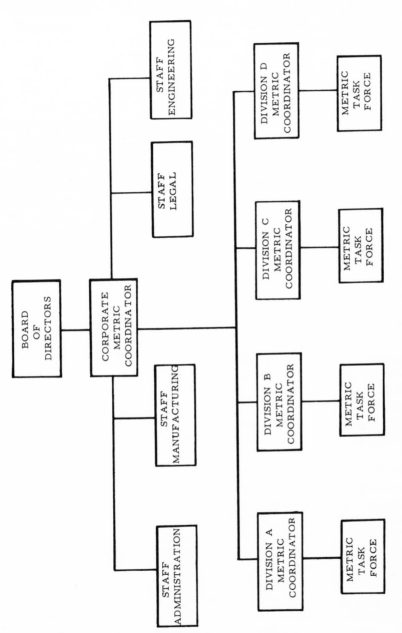

FIG. 6. Corporate metric organization chart.

27

In large corporations with several divisions, a decision must be made whether or not each division can convert to the metric system on its own plan and establish its own timetable. In particular, if overseas divisions are involved, metric will already be the predominant system of measurement. However, the SI metric system may not be used, so care in the transition planning is critical.

Committee responsibilities in a centralized corporation parallel those of the division-level committee in the multinational, decentralized corporation. In some instances, the board of directors will appoint a metric task force to investigate metrication and make recommendations. If the firm was found sensitive to metrication, the same committee would develop and implement a conversion plan.

Finally, the metric committee has the option of considering outside assistance for the metrication effort. There are a number of qualified metric consultants, but care must be taken to choose only those who are qualified to help your company. Some consultants might be experts in only one area of metrication, so it is important to select consultants on the basis of your company's needs.

Metric Coordinator at the Operating Division

Divisional metric coordinators should be responsible for planning and implementing metrication at the division level. They should be chosen by division management and, like their counterpart at the corporate level, should be high in the company organization and have the respect and cooperation of their associates. A division metric task force, comprised of the division coordinator and department heads involved in the conversion, should also be formed.

Finally, the coordinator's duties include:

1. Setting up committee meetings
2. Establishing departmental coordination
3. Initiating the division metric library
4. Serving as liaison between the division and the corporate office

Generating Metric Awareness

Changing to a new system of weights and measures will cause many psychological problems which must be resolved if metrication is to be successful.

Initially, the metric committee must be convinced of the benefits of metrication, and should be aware of current metric developments. Once the committee is knowledgeable, key people should be involved in the planning. A good method to do this is to establish a schedule requesting various department heads or representatives to join the committee meetings to discuss metric impact in their area. A metric educational briefing could be included to calm any transition fears. Out of this will come a better understanding of metric by more people and a feeling of participation in the transition process.

Once the involvement has begun, department heads should have metrication brought to their attention on a regular basis. This can be done in several ways:

1. Join the American National Metric Council and circulate their publications
2. Subscribe to metric periodicals and circulate them
3. Contact trade organizations on a regular basis to obtain the latest information and circulate this

There is an ever-increasing amount of metric information available from metric groups, trade associations, and the U.S. Government. A company metric library or information center would help to organize these materials so that they can be put to the best possible use.

Developing A Plan

The first thing to remember when developing a metric plan is that your company should not convert just for the sake of conversion. Metrication should take place only when it is economically feasible or when the marketplace will support it.

Metrication will usually go through three phases: (1) an investigation phase, where data are assembled and the metric committee educates itself in metric; (2) a planning phase, where the data are sifted and weighed and projections are made for implementation; and (3) an implementation phase (see Fig. 7).

Once corporate management has decided to go metric, a decision on the type of conversion must be made; i.e., soft or hard. Soft conversion involves changing numbers from standard to metric. There is no physical change in the product itself. Hard conversion indicates a change in the product itself. Consider impact on customers in this country and overseas. Consider impact on competition: would you gain a marketing edge? How would present

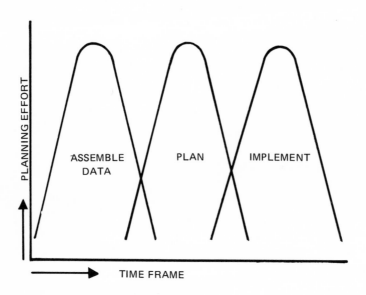

FIG. 7.   The three phases of metrication.

tooling and machinery be affected if you undertook hard conversion? The useful life of your product may affect your decision, as well as inventory on hand. Gather all the data before making the decision. Many elements affect the timing of metrication. For example, a large, multinational firm may be able to act as a leader in conversion; suppliers and customers will follow its direction. However, timing should be such that suppliers can make the necessary production changes. The best time to implement metric conversion may be when there is a design change or new products are introduced. The timing for the initiation of your whole plan may rest with your customers. If your firm does not produce an end product, you must contact your customers to establish guidelines to introduce metric units to them. Questions of competitor and market reaction will also influence your plans. Could customers use metric products now if they were available? The purchasing department must contact your suppliers to determine the availability of metric components and parts.

Once timing is determined, an overall conversion schedule can be established. Check points should be added to the schedule to monitor company metric progress. These points will vary depending on your products, but should include:

1. Training periods
2. Engineer drawings to metric
3. Product introduction
4. New packaging
5. Sales literature
6. Computer conversion
7. Machinery and inspection gages available
8. Metric inventory availability
9. Production target dates
10. Department conversion deadlines

The next phase of metric planning should include an analysis of your organization. Prepare an organization chart for your firm and a checklist for each department outlining what must be done. As mentioned in the section on generating metric awareness, personnel from other departments should be involved. This will assist the task force in writing checklists and educate others as well.

Part of the plan may include a pilot program. Use of this approach depends on the corporate structure, but a pilot program could be a valuable tool to determine conversion problem areas. The program could be set up as follows: (1) isolate a particular product; (2) educate the engineers, staff, foremen, machine operators, and all personnel involved; (3) purchase the necessary gages and equipment to produce and inspect metric units; (4) run the program for a specified period of time. A pilot program offers a unique opportunity to test your educational programs, production procedures, inventory ideas, and conversion plans on a small scale.

The legal aspects of metric may also affect your plans. Examine all contracts to determine implications. Vendor contracts may have to be updated, distributor and sales contracts may have to be altered, and labor agreements demand consideration.

The cost of conversion and training will be discussed in depth later, but they must be considered in the planning stage. It is strongly recommended that training be given to personnel near the time they will make use of it, and training should involve only what is required to do the particular job.

Finally, the plan should be flexible. International committees will continue to work toward agreement on more and more metric standards, and government legislation will have a distinct impact on the conversion timetable. Regular meetings should be held to keep abreast of standards revision and legislative action, and to review customer and vendor progress.

Implementing the Plan

When the plan is completed, it will be obvious that work can be started in several areas without any great expenditure. Before rumors begin to weave their way into your plant, consider publishing a progress report to keep employees from becoming overly concerned. There are advantages and disadvantages to doing this, but if this approach is adopted, the following method should be considered:

1. Meet with your department heads to explain the plan in depth and outline departmental responsibilities. Use the history of the metric system and the present disadvantages of the customary system to substantiate the need for change. Use your firm's position in the marketplace and other facts which provide a strong case for going metric.
2. Meet with your labor union representatives. Fill them in so they can answer questions brought up by other personnel. Be sure to discuss the length of transition and what will be expected from them. If a pilot program is in your plan, discuss this in depth.
3. Use some vehicle to contact your personnel. This can be accomplished through an employee newsletter or meetings to pass on the information.

Now begin setting up the transition. There is much that can be done without investing a great deal of funds. Develop a training program. As mentioned earlier, this may involve several different programs. Rewrite your engineering standards and contracts. Continue to monitor customers and vendors, and meet regularly to review the plan. Continue to evaluate your firm's position in the marketplace.

A point will be reached where a "proceed or hold" decision has to be made. If your firm is multinational and decentralized, the advantages of conversion may far outweigh any disadvantages. The decision was made before planning was undertaken: proceed with no holds. For many smaller companies, however, implementation can proceed to a point of committing to metric drawings, metric packaging, and metric tooling and gages. These firms may have to adopt a wait-and-see attitude until it is feasible to continue.

Implementation beyond this stage begins with Engineering, and new design drawings might be the best place to begin conversion. As the drawings proceed through the company, metric knowledge and awareness should increase. The activities of all departments involved should be coordinated to ensure that (1) gages are available when needed; (2) new packaging is ready; (3) space for metric inventory is available; (4) sales literature is ready for distribution; and (5) computer programs have been converted.

Chapter 3

ENGINEERING METRICATION

JOSEPH D. GARCIA

Corporate Materials Management
Combustion Engineering, Inc.
Windsor, Connecticut

The first task for an engineering department in a metric conversion program is to qualify and quantify specific areas of metric impact. Once this is completed, metrication can be planned and implemented in a manner best suited to yield positive results. Although identifying areas of metric impact may seem easy enough, many details must be considered:

1. What will be the impact on new and existing designs?
2. How will product lifetime affect metrication plans?
3. Should new standards be developed for
    a. design graphics?
    b. general engineering specifications?
    c. quality assurance specifications?
    d. engineering tables and data sheets?
    e. standard calculation procedures? (computer based or other)?
4. Should metric product design be
    a. pure SI?
    b. hybrid, metric/inch?
    c. inch module, soft converted?
5. What about
    a. rounding?
    b. dual dimensioning?

    c. codes and standards from outside sources?
    d. training?

One approach to answering the questions follows.

    The overall corporate metric organization is shown in Fig. 8. The engineering organization shown in Fig. 9 shows only the major segments of the organization; actually, each segment has a task force in which the disciplines to be addressed are represented, (i.e., heat and work transfer, mechanical, electronic, etc.) Thus, means for finding the metric impact on all levels of engineering are provided. Task assignments are designed to generate reports stating what is needed to operate in the SI metric system and how many man-hours it will take to achieve that posture.

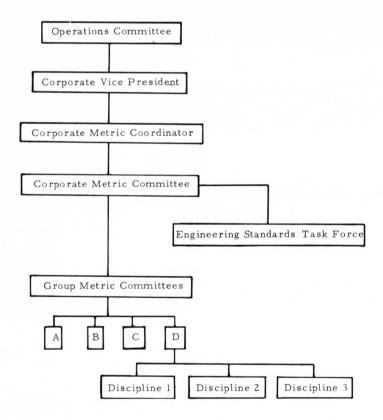

FIG. 8. Corporate metrication organization chart.

Power Systems Engineering

Research

Metallurgical & Materials Laboratory

Kreisinger Development Laboratory

Energy Systems Evaluation

Engineering Science

Product Development

Product Design & Performance Development

Performance Standards

Industrial Products Development

Product Systems Analysis

Product Design

Proposition Engineering

Performance Design

International Engineering

Product Engineering

Design Graphics

Systems & Component Design

Engineering Quality Assurance

Canada Engineering

FIG. 9.   Engineering metrication organization.

Levels of Activity

The data indicate three levels of probable activity:

1. A common level
2. An ongoing level
3. A final implementation level

The Common Level of Activity

In order to establish a uniform approach in deciding which SI units are to be used in engineering, a base reference publication should be selected for company use, such as ASTM E-380. Other common levels become visible as metrication progresses.*

Ongoing Activity

Ongoing activity includes addressing questions that have been previously stated. Addressing answers should reflect overall corporate strategy.

Let us first turn to the maintenance of engineering standards and documents. Engineering documents may be divided into three general categories:

1. General engineering specifications:
   For raw materials, mechanical components, fluid components, and electronic/electrical components
2. Project engineering specifications:
   Specific product application documents, such as customer requirements which deviate from standard
3. Quality assurance specifications:
   Interpretation of codes and regulations impacting on design, manufacturing, installation, and service
   These may subdivided into foreign and domestic. Examples are:
   ASME codes
   ASTM codes
   International Standards

*What may be common to one organization may not be to another. A selection of SI units "usable" by all engineering sections seems to be the logical starting place. Engineering policy should reflect consistency in application of all SI units.

Standards may also be subdivided into three categories:

1. Engineering tables, such as heat-transfer coefficient tables.*
2. Calculations: dimensional analysis of engineering calculations ensures the integrity of SI equivalency in commonly used formulas. *Note:* One very useful chart has been published by Netherlands Normalistic-Instituut, *Quantities and SI Units,* which is available from the American National Standards Institute for a small fee.
3. Design graphics: changes impacting this area occur in (among others):
    Drawing formats   Symbols
    Diagrams   Dimensioning

Here the opportunity presents itself to eliminate nonuniform practices in the development of standards. One approach which has been useful in this area is the use of the Pareto maldistribution curve. When applied to standards and documents, the graph is made up of "usage" vis-a-vis "number or volume" of standards. This provides a means for determining a starting place and focuses on overall priorities.

Discussion of the questions previously posed will, hopefully, yield some of the answers required for implementation. Subjects include dual dimensioning, rounding, codes and standards, product lifetime, and training.

Dual Dimensioning

The question of dual dimensioning will cause a variety of input from every engineering quarter. Dual dimensioning is taken for granted as being a part of metrication. However, we shall explore it further.

Dual dimensioning means the placement of two sets of related physical quantities expressed with different symbols on an engineering document. What is this really meant to accomplish? The engineer and/or draftsman has more to write. Even if the process is automated and the task is shifted to a clerical function, there still has to be a good reason to dual-dimension. Some observers state that it would make it easier for shop personnel to follow the change and, in fact, it would serve to train them. If you were in a machinist's shoes, would you use a metric unit if an inch unit was placed next to it or within easy reach? There are some situations that may necessitate dual dimensioning. An example may be when production is not to take place "in house." Design integrity may suffer if dimensional changes are not made "in house." Experience indicates that SI metric units should be used on new designs unless there is justification for doing otherwise.

*Where BTu/(h·ft² ·°F) may be shown as W/(m² · k) in SI notation.

Rounding

Converting to metric units will require procedures for rounding. The necessary accuracy of a calculation or a toleranced dimension depends on the specific circumstance in which it is applied; therefore, any rules for rounding must be applied when the individual situation is suitable. A simple approach is:

> Once the last significant figure has been selected and the digit that follows is less than 5, the last significant figure remains unchanged.

> *Example:* 2.3135 rounded to the second digit following the decimal point becomes 2.31.

> When the digits that follow the last significant figure are greater than 5 (this includes 5 followed by one or more quantitative digits) the last significant figure is rounded up.

> *Example:* 2.3151 rounded to the second digit following the decimal point becomes 2.32.

> When the digit that follows the last significant figure is exactly 5 and the last significant figure is odd, then and only then, the last significant figure is rounded up.

> *Example:* 2.315 and 2.305 rounded to the second digit following the decimal point become 2.32 and 2.30 because zero is considered even.

On the average, this yields as many figures rounded up as figures rounded down. Rounding to the nearest even number makes the arithmetic a little easier as well.

A full discussion of rounding is given in ASTM E-380 (ANSI Z210.1).

Codes and Standards

Codes and standards were previously mentioned in the section outlining engineering documents. The discussion will now focus on some of the items that may affect engineering strategy for metrication.

If engineering has, for example, selected kPa (kilopascals) as the unit for pressure, it will have to be changed to bars if product description is utilized in Germany. Similar problems exist when using Celsius in France.

Although the International Standards Organization (ISO) has attempted to represent an international approach to solving these problems, standards issued represent a generic approach to standardization. The specifics are still to be found in the individual countries.

Engineering must reflect its preferences and determinations through involvement *now* and not wait until metrication is completed; otherwise, we may find that other authorities may have written our standards for us. Metrication cannot afford to wait for production of standards and codes before being implemented. Involvement through participation and monitoring of codes and standards in the marketplace is an important function of engineering.

Product Lifetime

Let us assume that a new product is going to be introduced and that the lifetime of the product is anticipated to extend into the metric era. What considerations should be addressed?

Material

Will materials that have been made to inch modules be replaceable in a metric era? If not, can something be done? The decision must be made on specific metric product designs.

Metric Product Design

Pure SI Metric—a design which is completely based on metric modules and whose qualitative and quantitative representation is shown in SI metric notation

Hybrid—a design which is based on SI metric and inch modules, whose qualitative and quantitative notation is shown in either SI metric and/or inch notation

Inch Module—a design which is completely based on inch modules and whose qualitative and quantitative notation is an equivalent or soft conversion of the inch module written in metric notation

How should the new product be designed with respect to these possibilities? First, a metric product is wanted. Second, how is one to go about designing that product? Third, are the resources to produce the product available today?

Let us assume that the homework has been done and engineering has the capacity to begin a metric design. The decisions, then, concern a product designed now, but whose production lifetime will extend into the metric era. The object becomes to produce a design that works in both eras.

Will the pure SI metric design work under these constraints? Yes! But can a pure SI product be produced now? One answer lies with the availability of metric components. There are many component parts used in products which are not available in metric modules. We will consider these as being mating components.

A mating component may be standardized to the point where industrial standards exist [an example is a standard produced by the National Electrical Manufacturers Association (NEMA)]. Although it is not metrically based, knowledge of what the product may be like in the future should affect present design. However, now the mating component must be used, thus yielding a hybrid product.

The argument against dual dimensioning was presented earlier. However, let us address our metric product which is entering the inch world. Let us further assume that it is an all-world product. The product can be designed for production and usage out of the United States or, if produced in the United States where market research has said it would be detrimental to show SI metric notation, the design could be produced initially with dual dimensions and supplied to production with a drawing whose inch unit only would be exposed. In this manner, a nonvisible SI metric modular design can be introduced to users that are not yet ready for metric. Further, when the time is right for the metric product, all that need be done is to expose the metric notation and cover the inch notation.

This is one approach to having the best of both worlds, realizing that some time in the future, as more metric mating components become available, the product will evolve further into an all-metric state.

Training

On all engineering levels, metrication training begins very early by way of planning. The first comment one is likely to hear when metrication training is mentioned to a chief engineer is: "We can design anything you want, be it in inch or metric." On the other end of the spectrum another chief engineer may say, "It is a difficult task. A designer can now imagine, and be familiar with, something like pressure. He has a 'feel' for pressure, say on the palm of his hand, but how does he relate to force in newtons on that same area? How does he relate to pascals?"

If we say both of the above statements are true, then the focus is on an approach to training that will satisfy both conditions and any in between. To find out, however, there must be a means for testing the validity and extent of truth in both statements which will relate to the training one has in mind.

Consider this approach:

1. Training is to be specific.
2. Training is to be given just before a person will be engaged in metric work.
3. Individuals must receive training which will enable them to do their job.
4. Training material will be divided into small programmed modules.
5. Each training module will concentrate on a specific engineering discipline.
6. Means for determining which modules an engineering person needs can be developed through testing.
7. Testing means (can be likened to a sifting screen) will be directed at determining which general level or categories of study a person falls into and which modules within that level he or she needs.
8. Achievement recognition means is built into each module.

This approach fits both situations cited previously by the two chief engineers and provides a working solution.

Implementation

Strategy for implementing the SI metric units and quantities in engineering is based on overall corporate strategy. Achieving an early metric posture may mean the difference between a smooth transition vis-a-vis a panic-based one. It can be measured in terms of engineering doing its homework. Provision for the proper macroenvironment, by early planning in metrication, in such a way that the microenvironment can get its job done, is certainly the way to successful engineering metrication.

Implementing is nothing more than doing what one has been well prepared to do.

Chapter 4

MANUFACTURING METRICATION

JASSIE MASTER
Corporate Engineering Department
Ingersoll-Rand Company
Princeton, New Jersey

Management

The impact of metrication will be far-reaching. Every aspect of the manufacturing function will be directly affected, not only by the company's metrication policy, but also by the policies of its suppliers and subcontractors. The problems presented by metrication in a manufacturing organization will vary considerably with the precise nature of the product and the manufacturing processes involved.

Training and Communication

The implementation of metrication programs will require the very careful coordination of all departments within manufacturing as well as very close liaison with other functional areas, e.g., engineering, marketing, training, etc.

While it is recommended that some form of general appreciation training be given to all employees, it is considered that most practical training will take place on the job. Therefore, manufacturing managers must ensure that line supervisors are adequately briefed and, where necessary, themselves specifically trained to impart on-the-job training in metrication techniques.

Timing

The timing of the changeover to metric within manufacturing departments is a function of the company's overall metrication program. In addition, it is dependent on the availability of supplies, materials, machinery, etc. It is recommended that (1) a network analysis be prepared, planning the change-over to metrication; and (2) joint discussions between design engineering, manufacturing, purchasing, and quality control departments take place at an early stage in the development of a metric product. This is required so that sufficient lead time is allowed for the acquisition of necessary machines, equipment, and tooling and inspection instruments, and for training in the use of this equipment.

Costs

Metrication costs will have to be kept in mind when preparing manufacturing budgets. Principal areas of costs will be the conversion of machinery, equipment, tools, training of labor, the increase in storage space, double stocking of customary and metric materials and components, and the increasing cost of supplies in customary units as metrication proceeds.

Additional costs will arise due to the necessity to revise product planning and cost data in metric terms, and to rewrite test and inspection programs and specifications.

Rationalization

While the cost of metrication will be considerable, one of its by-products will be the opportunity to rationalize, not only by the strict views of preferred sizes of materials and components, but by the sweeping aside of many outdated methods and procedures and the introduction of modern production techniques and methods.

With careful planning, metrication can be used as the opportunity to introduce changes which will result in reduced production and operating costs.

Identification

It may be necessary to clearly identify, by color coding or other means, metric components, tools, machines, etc.

Industrial Engineering

The impact of metrication could impose an additional workload on industrial engineering personnel, and the metrication timescale should be arranged so as to avoid unnecessary overloads.

Planning and Estimating

Metrication may involve the rewriting and, in some cases, the rethinking of planning sheets and routine instructions. Many changes will result from the use of materials such as bars and sheets in metric stock sizes where machining and cutting allowances are different.

Departments such as sheet metal working and fabrication will face problems. Unless their operations are replanned and new cutting instructions produced, considerable material wastage and increased cost could result.

A complete new schedule of standards will have to be compiled. Where estimating or synthetic manuals are in use, conversion and rewriting will be required.

Product Tooling

A survey of subcontractor's plans should be carried out to ensure that metric requirements are not placed on firms whose facilities are inadequate.

New tooling for new metric products should be produced to metric standards. Very close liaison is required with any subcontractors producing components from a division's own tooling and with subcontractors producing tooling from which components will be produced.

Plant Maintenance

Obviously there will be the need to install new machines and to convert existing machines.

In addition, there will be the necessity to replace some maintenance tools and equipment, and to acquire additional metric small tools and measuring equipment.

Factory services will be measured and metered in metric units, and the conversion and recalibration of metering and indicating equipment must be considered.

All weighing and lifting equipment will eventually have to be recalibrated in metric terms, and the factory regulations will require provisions as appropriate.

Machines and Equipment

It is essential to plan future machinery requirements and replacements over a period of several years.

New Machines

In general, any acquisition of new machines at this stage of metrication should be the type having dual readout facilities or be capable of converting to dual readout. At an advanced stage in the metrication program, new machine acquisitions will preferably be restricted to machines with metric capability only.

Where new plants have to be purchased specifically for the continued production of nonmetric products, it is recommended that only plants which manufacturers are willing to guarantee future conversion to metric should be purchased.

Machine Conversion

When considering whether or not to convert existing machines, the life expectancy of the machine in its existing form should be closely considered. Conversion costs will not be incurred if the machine is scheduled to be replaced in a relatively short time. In considering whether to convert or replace, the following factors should be taken into account:

1. Delivery time for new machines
2. Delivery time for conversion equipment and time taken for conversion
3. Life expectancy following conversion

Where a machine is to be converted, a survey should be carried out for each separate piece of machinery and equipment to identify exactly which of the capabilities it possesses are, in fact, used in current manufacture. For most machine tools, a range of metric conversion kits is available. It is advisable to consult the manufacturers to ascertain availability, price, and source of these items.

In many machine tools, particularly those which are required to machine to a high degree of precision, difficulty can be experienced following conversion to metric due to wear in components other than those being replaced as part of the conversion. It is recommended that these aspects be carefully considered before embarking on a costly conversion program.

Investigation should be conducted on different devices available in the market to assure that the appropriate device (one that meets all the needs) is selected. Where dual readouts are added, the conversion equipment should incorporate a means of obscuring the units not applicable and of identifying the units to be used.

Machine Accessories

Although it will ultimately be necessary to replace machine accessories (for example, collets and arbors) with metric equipment, it is important to realize that the time of change has to be associated with the availability of raw materials in the metric sizes. Initially, it will be prudent to machine metric components from inch stock. It is at the time of appropriate changeover to metric standards for raw materials and stocks that most machine accessories will need to be replaced.

Small Tools

An opportunity exists to standardize on preferred metric sizes and the specification of metric specials is deprecated. Small tools are normally consumerable, but during the changeover period the economies of double stocking and additional storage costs need to be investigated.

Quality Control

Inspection and test specifications will, of course, be written in metric terms for new metric components and products. Where existing products are redesigned for metric production, inspection and test specifications will need to be rewritten, and any special test equipment will have to be redesigned or modified.

Inspection Standards

Manufacturing departments having a standards room will have to obtain new metric standards. It will no doubt be necessary to maintain both customary and metric standards for some time to come.

Inspection Equipment

The requirements for inspection equipment should be considered at the same time as the purchase of the machine tools which will produce the metric products.

In general, the comments already made regarding plant and equipment, its replacement and conversion, will also apply to inspection equipment.

Some inch measuring and gaging equipment can be converted to metric by recalibration, e.g., air gages and mercury gages. Other items of measuring equipment may be converted by the substitution of certain component parts. The advice of the manufacturers should be sought before expensive equipment is replaced. However, the life expectancy of the equipment should be considered. In general, a life expectancy of several years should be foreseen before conversion is undertaken.

Conversion Tables

Metrication must be so planned that machine equipment and measuring equipment using the units specified on the drawings are always available, both on the shop floor and in inspection departments. Conversion from inch to metric or vice versa must not be left to the operator or inspector. For these reasons, the use of conversion tables on the shop floor and in the inspection department should be discouraged.

Purchasing Stores and Stock Control

For some years, it will be necessary to carry stocks of both customary and metric items, and it is suggested that during the transition period some clear means of metric identification be developed.

Some customary items will become obsolete as a result of metrication. It may be prudent to maintain stock valuation accounts separately for customary and metric stocks so that the extent of stockholding of customary items can be closely monitored during the transition period.

Purchasing

Because many components and raw materials are subject to long delivery, purchasing departments must be brought into the metrication policy decisions from the beginning if continuity of supplies is to be maintained. This is particularly important as the effect of metrication is likely to increase delivery times rather than shorten them. Initially, there will probably be a general shortage of many items.

The buyer should be given firm planning dates for the introduction of metric products to allow for sufficient lead time:

1. To identify stocks of each item and raw materials to meet the program
2. To locate and evaluate new sources of supply
3. To arrange for new tools to be ordered, made, and tested
4. To secure batch quantities

Buyers must progressively obtain supplier's and subcontractor's programs, dates for conversion of machine tools, availability of materials, etc., and ensure well in advance of the introduction of new metric products that suppliers and subcontractors can meet their commitments.

Buyers should continuously obtain and maintain lists of all purchased materials, both in components and fastenings to metric standards, and ensure that at all times copies of these standards are passed to engineering, manufacturing, and other interested departments.

Stock Control

Stocks of inch items must not at any time exceed the quantities required to complete existing contracts, or authorized production runs for existing products. It is realized that some existing products may remain in use for several years.

Obsolete and surplus slow-moving items should be disposed of with the least possible delay to make room for the intake of metric components and materials.

Stores

A number of important decisions have to be made:

1. Total segregation of metric work from inch work, which will almost inevitably require additional storage facilities
2. Complete segregation of raw materials, components, and standard commercial items in all stores and work in progress
3. Purchase of storage equipment, bins, handling facilities, etc.; opportunities must be taken to effect a general improvement in stores efficiency at this time
4. Clear and uniform identification of all metric components, particularly in those cases where a metric item is not obviously different from its inch counterpart
5. Metric measuring and weighing equipment in goods received departments to be ordered and installed in advance of the receipt of metric supplies

Chapter 5

METRICATION EFFECT ON MARKETING

R. G. SLOAN AND ARTHUR M. LENNIE

New Product Planning
Armco Steel Corporation
Middletown, Ohio

Marketing is the performance of those business activities that direct the flow of goods from producer to consumer or user. We have begun with a fundamental definition because metrication is affecting such a broad scope of industrial activity, because the term marketing can have many connotations, and because marketing is structured into corporate organizations in many ways.

Accordingly, we will be describing the interaction of metrication and, in a broad definition, the entire commercial end of business. What is more, with the variety of industries involved in metrication and the range of companies in any given industry, it is understandable that any discussion involving marketing must be restricted to general principles and guidelines. This discussion must necessarily be so restricted because our marketing and metrication experiences are primarily limited to the steel industry. But the basic principles of marketing apply whether you are involved with consumer soft goods, pumps and motors, or steel. And we are thoroughly convinced that the same situation applies to metrication.

Certain principles apply to metric conversion or adoption in *all* companies in *any* industry. However, the implementation of metrication and its effect on marketing will undoubtedly not be the same in any two companies in any industry. Reasons for industry differences are obvious; those for the differences from company to company not necessarily so. It is to the end of evolving logical and effective means of determining those differences that we must address ourselves.

Key Marketing Aspects

Before attempting this, we feel it is necessary to review some fundamental aspects of metrication as they relate to marketing.

The first is that commercial or marketing facts and judgments are decisive factors in metrication decisions, from the establishment of a company's basic policy to its detailed planning and implementation. This fact tends to become overlooked as metrication progresses at an ever-increasing pace.

Marketing often receives less than required attention because metrication literature features the more intricate and detailed problems of how to make the conversion. This is understandable because the leaders in metrication are large and usually multinational companies. They have, with the assistance of sophisticated marketing staffs, thoroughly analyzed the commercial impacts of metrication. And, by nature of their international operations, they have opportunities to most effectively utilize resultant efficiencies and economies. Note the attention given to commercial aspects in the following selected quotations from explanations of why companies such as Ford, GM, IBM, and International Harvester are converting to SI:

Product uniformity enhances marketability.

Based on realistic appraisal of products and markets.

Metricating to improve efficiency and international commerce.

Worldwide activities require metrication. It will be based on need and economic feasibility.

Plans are based on marketplace needs, opportunities to minimize costs and increase market share.

Each division will progress as needed, based on customer requirements and anticipated markets.

Moving on metrication to protect our worldwide industrial position.

Metric plan will be timed primarily in accordance with our customer requirements.

If marketing influenced the decisions on metrication in multinational companies that can most easily exploit the advantages of metrics, marketing must necessarily be more decisive in smaller organizations operating in more restricted areas. In essence, particularly for the latter, marketing will impact metrication as much or more than metrication will affect the marketing function.

The second fundamental aspect is to clearly understand the full import of the three basic questions on metrication: (1) Whether to?, (2) When?, and (3) How?

The first question is probably academic for most industry in the U.S., inasmuch as we seem to be well on our way to metrication. Major segments of key industries have made the decision, and their decision answers the question immediately for many other companies. For example, when the automobile industry indicated that it would be producing metric units and would need metric-dimensioned steel, the steel industry prepared to produce it.

With conversion well underway in so many companies and given the imminence of congressional action, the only logical answer seems to be that, sooner or later, we will all convert.

The second question (when?) then becomes the most important. This resolves into decisions of timing in serving your customers most effectively and optimizing your company's profit. This question, like the first, is easily resolved if your major accounts specify when they want metric products.

This easy answer will not apply to most enterprises. They will have to ask themselves some hard questions: How sensitive are our customers and markets to metrication? What are their conversion plans? How will the voluntary aspect of U.S. metrication affect conversion in our markets? Marketing must provide accurate input on which to base this management decision; generalities or metrication platitudes can misdirect the timing decision, resulting in increased conversion costs and marketing problems. Each company must resolve this problem based on its unique position. Your timing may range from immediate action on an all-out program to no activity at all for a period of years. However, whatever action is taken must be based on logical marketing analysis. Just sitting on your hands with a wait-and-see attitude might find you totally unprepared for the wave of metrication. Even if you take no immediate action, make sure your decision is a part of an overall metrication plan.

The last metrication question (how?) is primarily a technical one and other sections of this publication are devoted to providing suggestions in this area.

The third fundamental aspect of metrication as related to marketing is the fact that metrication in the U.S. will be a voluntary conversion. It is especially important to recognize the impact of "voluntary" on the conversion process. Essentially, voluntary action insures a market-driven change and makes the timing of your action critical. This voluntary emphasis in the U.S. is somewhat in contrast to the conversions in Britain, Australia, New Zealand, Canada, and South Africa. All of them, even when stipulated as voluntary, have either strong government direction or mandatory provisions which change the name of the game. This is not to say that nothing can be learned from their experiences. To the contrary, valuable information is being provided, particularly in the area of how to accomplish conversion. But when government controls major segments of markets, such as the construction industry in Britain, or stipulates the "when" of conversion as in Australia and South Africa, you are not operating under voluntary action. The point of emphasis here is that the voluntary aspect of our conversion process increases the effect of marketing on metrication.

Inasmuch as marketing is such a vital force in all aspects of the metrication process, it is our opinion that marketing or commercial officials must be key members, if not the directors, of company metrication committees. There are two reasons for this opinion. One is that, as we have discussed, marketing plays such a critical part in metrication decisions on policymaking and planning. The other is that marketing managers must not only provide essential input at the planning stage, but must also be intimately involved throughout the implementation to provide continuing market information and to formulate effective marketing plans.

Subscribing to the importance of strong commercial representation, Armco structures its corporate metrication committee as follows:

Chairman:	Senior Vice President, Commercial
Coordinator:	Director, New Product Planning and Technical Services
Secretary:	Manager, New Product Planning
Members:	Director, Sales Administration
	Director, Sales Planning
	Supervisor, Sales Order Entry
	Assistant Vice President, Engineering
	Director, Quality and Service
	Director, Industrial and Systems Engineering
	Legal Counsel
	Steel Plant Coordinators (6)
	Divisional Unit Coordinators (4)
	Assistant Controller, Corporate
	Steel Group Controller
	Research Administration Manager

Supplementing the corporate committee are committees in individual plants and special committees appointed to achieve specific objectives. In Armco's metric conversion, which has proceeded to the point of shipping our first orders of metric-dimensioned flat rolled steel, we have found the strong commercial-marketing composition of our committee invaluable. The reason is that most of the problems encountered and the decisions required were commercially rather than technically oriented. Our conversion efforts to date have been concentrated on flat rolled steel products, because our key customers indicated that the time is now. Additional phases of our business (other basic steel mill products, steel buildings, construction products, fasteners, oil well drilling and production equipment, and industrial products) are in various planning stages of conversion. But the timing of each will be dictated by the rule of reason — the dictate of the marketplace.

Orienting Metrication And Your Markets

Studying the effects of marketing on metrication is useful for guidance purposes. However, this will not get the conversion job done. Remember, we are dealing with specific products and specific markets that require specific plans and action. Orienting these products and markets to metrication is therefore essential.

The first step seems to be an objective review of metrication progress, both worldwide and in the U.S. An overview of where we are, how and why we got there, and how fast we are going gives you a "feel" for metrication.

While worldwide conversion activities are of interest, most of us will be primarily concerned with the status of metrication in the U.S. For both getting the "feel" and keeping abreast of current developments there is no better source of information than the American National Metric Council (ANMC). By participating in the Council, you will know (1) the status of legislation; (2) how various sectors of industry are reacting to and planning for metrication; and (3) what kinds of problems are being encountered and how they are being solved.

More detailed study of those markets and industries of primary importance to you is required. Here we must turn to the task of knowing precisely what is going on: what is the sensitivity of your key markets to metrication? Take the markets for flat rolled steel as an example. We know that automotive and farm industrial equipment manufacturers are converting rapidly and we monitor their activities through personal contacts. Similar investigations show that other steel markets, such as appliance and construction industries, are much less sensitive to metrication. All are being followed closely, however, because their sensitivity may change and we must be ready to serve that changing need.

This job is complicated by the fact that you are aiming at a moving target. The companies in your markets are facing the same problems; i.e., planning and timing their conversions. And their plans are not going to be static. But this task had best be done precisely or you may be led down a metrication garden path. For example, if a supplier of mill equipment or tools assumed that because Armco is producing metric steel, they are ready to purchase metric equipment, he could be absolutely wrong. The assumption is in error because we are soft-converting metric orders and our mills are producing in English units on our present equipment and will be for some years.

Our steel mill example may not be typical because it is simpler than situations which will be faced by companies producing goods for the consumer and manufacturing sectors. But the same principle applies. Know precisely how those key markets are moving in metrication, dig for information related to your products, and do not be misled by assumptions or generalities.

Cautions to avoid generalities make it axiomatic to orient your metrication toward that of your major customers. We are sure you will find, as we did at Armco, that it is essential to know in detail what those customers plan to do and when. To get such information, normal market intelligence and sales contacts may not be adequate, inasmuch as most companies, even if they have established a policy, are still in the process of converting. Here again, you have the problem of trying to hit a moving target.

If you are a supplier to the automotive industry, you are well aware that your customers are firmly committed to metrication. Their policies and general plans are known. But do you know specifically how the detailed plans affect your products? A review of ANMC Coordinating and Sector Committees reports indicates that many industrial sectors are still in a very fluid conversion situation. In some sectors, individual company actions range from passive interest to full-scale implementation. With a constantly changing picture, your customers' metrication plans may be revised tomorrow. But, fortunately, you are in the same boat. Also as we have said, you will probably find your customers as eager to know about your conversion plans as you are about knowing theirs.

Finally, in orienting metrication and your markets, it would be a good idea to keep an eye on your competitors' activities. In addition to the necessary collection of information, many judgments must be made to plan your metric marketing effectively. Since none of us is perfect, a comparative evaluation of what the other fellow is doing may prove fruitful.

While most of the foregoing discussion applies to the policymaking and planning phases of metric conversion, it also serves another purpose. Understanding the factors involved as well as preparing an adequate market analysis is essential to developing marketing plans that can effectively exploit your conversion to SI metric.

Armed with the required information, the marketing plan for metric conversion should encompass all phases of the conversion, from publicity on the formulation of a corporate policy to implementing technical service on your new metric products.

Complete involvement of marketing will uncover basic problems that might otherwise be overlooked. For example, production considerations might seem to indicate a product-driven conversion. However, market analysis could show that a market-driven conversion is preferable. Similarly, coordinating marketing and manufacturing analysis can more effectively determine whether a soft or hard conversion is preferable, or whether your product lines dictate that both methods need to be employed for some time.

Metrication offers the opportunity to simplify or revise your product line or packaging. While this is one of the major potential advantages of conversion, it requires a detailed coordination effort by manufacturing and marketing and can involve legal problems related to possible antitrust action. This problem will be discussed later in this chapter.

Given the strong emphasis on marketing, Armco has felt from the inception of its metric awareness that, in a voluntary conversion process, commercial policy must dictate. After study and analysis, this position has been confirmed in our corporate policy on metrication:

> It is the policy of Armco Steel Corporation to proceed with metrication as required to most effectively meet the needs of our customers. Implementation of this policy will concur in every practicable way with such policies as are established by the Federal Government, the American Iron and Steel Institute, and the industries Armco serves.
>
> Armco will supply metric-dimensioned products whenever possible within the limits of practical economics for both Armco and our customers. By mid-1975 it is Armco's intention to accept orders for flat rolled products in either customary or metric units. Metric-dimensioned orders for such products will be produced in equivalent customary units but invoices and shipping papers will be provided in metric or dual units, as required. Production of all other steel mill and fabricated products in metric standard sizes will be considered by Armco whenever orders for specific

products and sizes are large enough to enable the cost of necessary rolls, tool and dies to be amortized adequately.

Effect of Metrication on Marketing Functions

Up to this point, our discussion has dealt with the interaction of metrication and marketing, and has stressed the importance of commercial considerations in metric conversion. Let us now take a more specific look at the individual marketing functions and observe how metrication directly affects them.

In a broad consideration, you can make the assumption that metrication should not basically affect sales, market research, advertising, or any other marketing unit. This is based on the proposition that a metric product can be considered as simply a new or modified product. On this basis, all that is needed is development of a marketing plan to move the product to the consumer or end user effectively.

This admittedly amounts to oversimplification, inasmuch as there are a few unique marketing problems caused by metrication. However, it is a point to use in overcoming negative attitudes and reducing unnecessary complications. If you have not already encountered those who look on metrication as an unnecessary evil or envision all kinds of imaginary problems, you will. But an explanation of how simple the conversion process is can work wonders.

One of the impacts of metrication that affects all facets of marketing is the necessity of careful, detailed planning. Developing and carrying out any marketing plan requires precise planning. But in metrication, we are dealing with consumers or users of our product who are entering a new metrological world. This new world and all of its ramifications creates a unique situation, a situation which must be fully understood. This is why we have so strongly emphasized the necessity of studying all the basic aspects of conversion and their associated problems. Metrication does not really complicate marketing, but it does demand careful consideration of its uniqueness in the current marketplace.

Sales Function

The first assignment for field sales and sales planning is to let your accounts know precisely where you stand and what you plan to do. How and when this is done depends on the sensitivity of each market segment to metrication.

Pricing metric products in the materials and manufacturing sectors presents no basic problems. Unit prices of materials, whether in dollars or cents per kilogram, metric ton, or meter, require some judicious selection. But the business world should acclimate readily. At Armco, we sell steel in dollars per 100 pounds or cents per pound; metrically, the prices are dollars per 100 kilograms and cents per kilogram. Actual costs are the same in either system. We believe that the difference in the weight (mass) unit will be accepted with no problems. So far, this has been the case.

Consumer products sold in volume or mass units present a more difficult problem which relates to familiarity and acceptance of liters and kilograms and the resulting price differences. Movement is underway in making the change. But we will not be so presumptuous as to predicate on this problem.

One of the key advantages of metrication is rationalization, with the economic benefits to both producer and user. One of the most effective ways of achieving the use of standard metric sizes is to put a premium on nonstandard sizes. However, effective implementation of standardization on fewer sizes requires industrywide action. As you might expect, this activity exposes companies to possible antitrust liability.

The American National Standards Institute (ANSI) is establishing preferred sizes for metric products, but the participation by industry in furthering rationalization is limited by antitrust controls. It is hoped that congressional action will clarify this potential liability. However, in the interim, this is one sales activity to be approached carefully. Too many of the aspects of rationalization can be interpreted adversely.

ANMC has developed recommended procedures to be followed by industry groups that will minimize antitrust liability. However, companies will do well to keep their legal departments fully apprised of all metrication actions.

Once prices and price bases are established, what is the best way to present them to your customers? Practices will vary so widely in this respect that we are going to restrict this discussion to how Armco revised its pricebook and why.

First, the decision was made, even though metric sales would be limited for some time, to issue a separate metric pricebook. Pages were color-coded to distinguish them clearly. A metric price book was deemed necessary to prevent confusion in converting values. It is a soft conversion of our standard unit pricebook that provides all dimensions, limits, tolerances, etc., in equivalent metric units. We anticipate that a soft-converted metric pricebook will be used until, some years later, the majority of our orders are metric.

One of the most important effects of metrication on sales relates to communication with the customer in a new and different measurement language. By far the most important communication is his order for your product, and it is now going to be in SI metric. How do you handle the problem? You are not going to do it exactly the way Armco did, but a review of our procedure will provide guidelines that should prove useful.

What was our situation? We have a computerized ordering system and decided to soft-convert metric orders. That is, we would accept metric orders; convert these to English units; produce as English-dimensioned steel; and issue order, production, shipping, invoice, and record documents in metric and/or English units as required. Only flat rolled steel orders were involved, but this required coordination between corporate headquarters and five steel plants.

The first step was the formation of a task force consisting of representatives from sales, sales planning, order entry, sales service, systems engineering, production planning, metallurgical, shipping, traffic, and invoicing and sales records.

Because conversions of dimensional values would be made, the coordinator's office was established as the corporate source for all conversion factors, methods of rounding, and establishment of degrees of accuracy for all measurements. These were in accordance with standard procedure, but the central corporate source was established to insure uniformity and compliance with industry standards.

All documents relating to an order and records were reviewed by the task force. Those relating to a specific operation were analyzed in detail, and specific problems were resolved by appropriate members of the group. In addition to correlating the required input and output of these related documents, specific units to be used and significant figures for various measurements had to be determined, as well as whether metric, English, or both units were needed.

In some cases, consultation was required with customers to clarify some points. In more than one instance, our questions involved problems the customer had not yet considered.

Several months' work produced a proposed metric ordering system. But a dry run uncovered some minor errors. The system was then reviewed with key customers to insure communication compatibility. At this stage, we were able to show them that they needed both English and metric units on some documents that they had stipulated should have only metric units.

Previously, the ordering system had been involved in conjunction with only one of the five plants to be involved. This was done to simplify the basic work. Now all plants were evolved and further revisions of the system made to provide the flexibility necessary to adapt to their specific requirements.

With the system set, all inputs and outputs determined, the computer processing work was completed. The net result of this intensive and somewhat lengthy effort is an order processing system that meets the needs of our customers and resolves the metric impact on our most vital sales communication channel. What is more, we have established a basic procedure that can be utilized effectively as other units of Armco face the same problem. Despite the fact that we were working with a computerized system, our systems engineers assure us that the basic procedure can be applied equally well to manual systems.

Our ordering system, as we suspected, turned out to be the sales element most severely impacted by metrication. Resolution of this problem clearly emphasized to us the necessity of careful surveys to unearth problems and the absolute necessity of very careful analysis to eliminate all metric bugs.

A necessary adjunct to documentary communication was insurance that all other communications channels were converted. Sales personnel at headquarters and in field offices, as well as plant personnel involved with order handling, have received basic training in the fundamentals of SI metric as well as Armco's metrication plan.

Sales and Technical Service

While these two units of the marketing team are not affected in their basic function by metrication, metrication certainly increases their importance. Their combined responsibility is to keep the customer happy by helping to solve some of his day-to-day problems and assist him in using your product more effectively.

As you progress with metrication, your customers are going to have some problems with metric products, be they real or imagined. Even metric-oriented problems that are figments of a purchasing agent's or production supervisor's imagination are nonetheless real to him. In these cases, a knowledgeable service representative can stop trouble in its tracks as well as score an extra point for your team.

If your customers are metrication neophytes or hangers-back in conversion, use your metrication knowledge as a service tool. Not only can you give them valuable information to forestall the unreal problems, but you can also provide guidance for the real problems they undoubtedly will encounter. For example, your distributors or servicing centers can probably use help in training their people, and some help in inventory control will certainly be welcome. With manufacturing customers, technical service can be very useful in pointing out the advantages of standard sizes in both manufacturing and inventory control. These customers, especially smaller companies too, would also welcome help in training.

While metrication can be exploited as a useful means of providing good sales and technical service, it requires a very thorough and complete understanding of metrication by your service representatives.

Market Development and Market Research

The first sections of this chapter emphasized the importance of marketing intelligence in all phases of metrication. This in itself indicates one effect of conversion on these commercial arms of any company — they become more important than ever.

It would be redundant to say more, other than to emphasize the necessity that market research and development provide a continuing flow of market information. Metrication in the U.S. is in a very fluid state and will undoubtedly remain so for some time. In view of this, all metrication planners in marketing and elsewhere in the company are going to need accurate and current market information to guide their efforts in converting to SI metric most effectively.

Publicity, Advertising, and Promotion

The impact of metrication on these marketing units can best be described as a challenge to exploit metrication to the fullest advantage. There are opportunities to effectively use advertising and promotion publicity in your metrication marketing plan. How and when depends primarily upon your customers' sensitivity to metrication as well as the basic principles of advertising and promotion. But the newness and uniqueness of SI metrics do not abridge the rules of reason and good judgment.

However, these words, new and unique, indicate that publicity opportunities exist. For some time to come, depending upon the industry, metrication activities will be news. First, any significant developments in a new activity are always news. Armco, for example, got extensive publicity in trade publications and newspapers on both our corporate policy release and on our first shipment of metric steel, one of the first by the steel industry. If a coil of steel can get good coverage, so can your product. Opportunities for good publicity will undoubtedly increase when Congress enacts legislation on metrication. It is well to remember that good publicity never hurt anyone.

Part of the challenge metrication puts to advertising and promotion is to thoroughly understand metric conversion, especially what it means to your company, what it means to the customer, and how important it is to him. The extent to which your advertising and promotion people understand metrication will not only be reflected in your advertising per se, but even more so in your catalogs, technical manuals, and promotional efforts. Well-planned, accurate, helpful advertisements and literature can help you not only gain maximum advantage, but also can help you overcome any disadvantages of metrication that might arise.

One disadvantage can be problems with distributors or dealers, particularly in the beginning of your conversion program. Probably not too familiar with SI metric, they are the interface between your metric product and the consumer or user, who is even less familiar with metric. Also, your distributors, are going to be faced, at least temporarily, with inventory and service problems. For this reason, advertising and promotion people must be thoroughly metricated. Only then can they anticipate what the distributor will face, and produce appropriate local advertising aids, consumer brochures, service manuals, and other useful materials to supplement the efforts of sales and technical services.

As these examples illustrate, metrication poses a significant challenge to advertising and promotion. But have you ever met anyone in that end of the business who did not firmly believe that every challenge is an opportunity?

In Retrospect

To return to the beginning, marketing is the performance of those business activities that direct the flow of goods from producer to consumer or user.

Metrication is not only going to change the goods but inject a new metrological language. Ipso facto, metrication impacts marketing.

These are the facts we face, just as we face the fact that metrication is well on its way in the United States. But this presents no serious problem. An aggressive, capable marketing staff will have no difficulty in making the most out of the opportunities that metrication will offer your company.

Chapter 6

METRICATION OF COMPUTER SYSTEMS

JOHN L. KUHARIK

Department 540F
IBM Corporation
Endicott, New York

Before planning to convert computer operations can begin, a working knowledge of the SI metric system is necessary. Is there anything new about it? What are the basic ground rules for its use? The following explains the International Standards Organization (ISO) SI metric system and presents rules of metric practice as adopted by IBM.*

In the SI system, the seven SI base units of measure and supplementary units are:

SI Base Units

Unit Name	Plural Form	Pronunciation	Symbol	Quantity	Multiples and Submultiples
ampere	amperes	am′pi(ə)r	A	electric current	kA, mA, μA, nA, pA
candela	candelas	kăn de′lə	cd	luminous intensity	
kelvin	kelvins	kĕl′vᵉn	K	thermodynamic temperature*	mk
kilogram	kilograms	kĭlĭ′grăm	kg	mass	Mg, g, mg, μg
metre	metres	mē′tᵉr	m	length	km, cm, mm, μm, nm
mole	moles	mōl	mol	amount of substance	kmol, mmol, μmol
second	seconds	sek′ənd	s	time	ks, ms, μs, ns, ps

*Degree Celsius (°C) accepted (°C = K − 273.15). Plural form is degrees Celsius.

*This material is taken from the IBM publication, *SI Metric Style Manual.* Excerpts from another IBM publication, *SI Metric Reference Manual,* are also included in this chapter. Neither of these publications has been reviewed by ANMC's Metric Practice Committee, and therefore may not be in agreement with the recommendations presented in ANMC's *Metric Editorial Guide* (2nd edition).

SI Supplementary Units

Name	Plural	Pronunciation	Symbol	Quantity	Multiples and Submultiples
radian*	radian	rā´dē ən	rad	plane angle	mrad, μrad
steradian	steradian	stə rā´dē ən	sr	solid angle	msr, μsr

*Use of degree, minute, and second is acceptable

Only the units for temperature, mass, and length are new; the rest (electric current, luminous intensity, amount of substance, and time) are in common usage. Therefore, we will concern ourselves only with these units. Those who have had previous experience with the metric system and who study the SI system in depth will find that the obvious changes in the system are that basic derivation of a familiar unit has changed, as in the case of the meter (which we will not discuss because it is not pertinent to this study), or that a name has been changed, as in the case of temperature (degree Celsius instead of centigrade).

The prefixes which designate the change of the basic unit remain the same:

Prefixes of SI Units

Name	Pronunciation	Symbol	Amount	Multiples and Submultiples	Definition
tera	tĕr´á	T	1 000 000 000 000	10^{12}	one million million times
giga	ji´gá	G	1 000 000 000	10^{9}	one thousand million times
mega	mĕg´á	M	1 000 000	10^{6}	one million times
kilo	kĭl´ŏ	k	1 000	10^{3}	one thousand times
hecto	hĕk´tŏ	h*	100	10^{2}	one hundred times
deka	dĕk´á	da*	10	10	ten times
deci	dĕs´ĭ	d*	0.1	10^{-1}	one tenth of
centi	sĕn´tĭ	c*	0.01	10^{-2}	one hundredth of
milli	mĭl´i	m	0.001	10^{-3}	one thousandth of
micro	mī´krŏ	μ	0.000 001	10^{-6}	one millionth of
nano	năn´ŏ	n	0.000 000 001	10^{-9}	one thousandth millionth of
pico	pē´cŏ	p	0.000 000 000 001	10^{-12}	one millionth millionth of
femto	fĕm´tŏ	f	0.000 000 000 000 001	10^{-15}	one thousandth millionth millionth of
atto	ăt´tŏ	a	0.000 000 000 000 000 001	10^{-18}	one millionth millionth millionth of

*Avoid these whenever possible.

The derived units and symbols will be:

SI Derived Units With Special Names

Unit Name	Plural Form	Pronunciation	Symbol	Quantity	Formula	Multiples and Submultiples
coulomb	coulombs	kü′läm	C	electric charge	A·s	kC, μC, nC, pC
farad	farads	far′ad	F	electric capacitance	C/A	mF, μF, nF, pF
henry	henries	hen′re	H	inductance	Wb/A	mH, μH, nH, pH
hertz	hertz	hərts′	Hz	frequency	1/s or s^{-1}	THz, GHz, MHz, kHz
joule	joules	jü(ə)l′	J	energy	N·m	TJ, GJ, MJ, kJ, nJ
lumen	lumens	lüm′ən	lm	luminous flux	cd·sr	
lux	lux	laks′	lx	illuminance	m^{-2}·cd·sr	
newton	newtons	n(y)ü′t°n	N	force or weight	m·kg·s^{-2}	MN, kN, mN, μN
ohm	ohms	ōm	Ω	electric resistance	V/A	GΩ, MΩ, kΩ, mΩ, $\mu\Omega$
pascal	pascals	pas′kəl	Pa	pressure or stress	N/m^2	kPa
siemen	siemens	sē′mən	S	conductance	A/V	kS, nS, μS
tesla	teslas	tes′lä	T	magnetic flux density	Wb/m^2	mT, μT, nT
volt	volts	vōlt′	V	electric potential	W/A	MV, kV, mV, μV
watt	watts	wät′	W	power	J/s	μW, TW, GW, MW, kW, nW, mWb
weber	webers	web′ər	Wb	magnetic flux	V·s	

Some Derived Units Without Special Names

Unit Name	Symbol	Quantity
square metre	m^2	area
cubic metre	m^3	volume
metre per second	m/s	velocity - linear
radian per second	rad/s	angular velocity
metre per second squared	m/s^2	acceleration - linear
radian per second squared	rad/s^2	angular acceleration
newton metre	N·m	moment of force
kilogram per cubic metre	kg/m^3	density
joule per kelvin	J/K	entropy
watt per square metre	W/m^2	thermal flux density

In summary, the SI metric system, as agreed to by the member nations of ISO, is fundamentally the same as the old metric system. However, an important point to note here is that the symbols which have been shown are the proper SI symbols. There are acceptable alternate symbols as shown in Fig. 10.

Symbols - Alphabetic by Symbol

It is important that the symbols in italic type be italicized in use. This distinguishes them from other similar abbreviations and is essential to their correct usage.

The Alternate Symbols for Special Applications are to be used *only* in cases where the typewriter or printer does not have the capability of reproducing the symbols (as shown at the left) and it is not possible to spell out the word.

		Alternate Symbols for Special Applications	
Symbol	Definition	Using Uppercase	Using Lowercase
A	ampere	A	a
Å	ångström	ANG	ang
a	atto (10^{-18}) or are or year	A,ARE,ANN	a,are,ann
a_0	Bohr radius	A0	a0
AE	German usage, astronomical unit	AE	ae
atm	standard atmosphere	ATM	atm
AU	astronomical unit (the symbol for this unit is UA in French and AE in German)	AU	au
b	barn	BRN	brn
bar	bar	BAR	bar
C	coulomb	C	c
°C	degree Celsius	CEL	cel
c	speed of light in vacuum	CLIT	clit
c	centi (10^{-2})	C	c or d-2
cal	calorie	CAL	cal
cd	candela	CD	cd
cd/m²	candela per square metre (luminance)	CD/M2	cd/m2
Ci	curie	CI	ci
d	deci (10^{-1}) or day	D, D	d or d-1, d
da	*deka (10)	DA	da or dl
dB	decibel	DB	db
dyn	dyne	DYN	dyn
e	electron charge	E	e
e/m_e	electron charge to mass ratio	E/ME	e/me
erg	erg	ERG	erg
eV	electronvolt	EV	ev
F	farad	F	f
F	Faraday constant	FC	fc
°F	degree Fahrenheit	DEGF	degf
f	femto (10^{-15})	F	f or d-15
G	giga (10^9)	G	d9 or g
G	gravitational constant	GC	gc
g	gram or grade (angle)	G,GDE	g,gde
gal	gal	GAL	gal
g**	grade (angle)	GON	gon

*An alternate spelling, not used in IBM, is deca.
**Symbol is used in the right superscript position.

FIG. 10. Symbols — alphabetic by symbol.

H	henry	H	h
h	hecto (10^2) or hour	H,HR	h or d2,hr
h or $h/2\pi$	Planck's constant	PC	pc
$h/2m_e$ or h/m_e	quantity of circulation	H/2ME	h/2me
ha	hectare	HAR	har
Hz	hertz	HZ	hz
in	inch	IN	in
J	joule	J	j
K	kelvin	K	k
k	kilo (10^3)	K	k or d3
k	Boltzmann constant	BC-K	bc-k
kg	kilogram	KG	kg
kg/m^3	kilogram per cubic metre (density)	KG/M3	kg/m3
l	*litre	L	l
lm	lumen	LM	lm
lx	lux	LX	lx
M	mega (10^6)	MA	ma or d6
m	*metre of milli (10^{-3})	M,M	m,m or d-3
m^2	square metre (area)	M2	m2
m^3	cubic metre (volume)	M3	m3
m^3/kg	cubic metre per kilogram (specific density)	M3/KG	m3/kg
m_e or m_e/u	electron rest mass	ME or ME/U	me or me/u
min	minute (time)	MIN	min
min^{-1}	revolutions per minute	REV MIN-1	rev min-1
m_n or m_n/u	neutron rest mass	MN or MN/U	mn or mn/u
mol	mole	MOL	mol
mol/m^3	mole per cubic metre (concentartion of amount of substance)	MOL/M3	mol/m3
m_p or m_p/u	proton rest mass	MP or MP/U	mp or mp/u
m/s	metre per second (speed, velocity)	M/S	m/s
m/s^2	metre per second squared (acceleration)	M/S2	m/s2
N	newton	N	n
n	nano (10^{-9})	N	n or d-9
N_a	Avogadro constant	NA	na
Np	neper	NP	np
Nr	number	NO.	no.
P	poise	POI	poi
p	pico (10^{-12})	P	p or d-12
Pa	pascal	PA	pa
pc	parsec	PC	pc

*Alternate spellings, liter and meter are used in the U.S., however, these are not used in IBM.

FIG. 10. (cont.)

R	gas constant	RGC	rgc
R	roentgen	R	r
R^{∞}	Rydberg constant	RBG	rbg
rad,rd	radian, or rad (unit of radiation dosage - an alternate symbol, rd, may be used to avoid confusion)	RAD,RD	rad,rd
r_e	classical electron radius	RE	re
S	siemens	SIE	sie
s	second (time)	S	s
SI	Système International d'Unités or International System of Units	SI	si
sr	steradian	SR	sr
St	stokes	ST	st
T	tesla and tera (10^{12})	T,T	t, d12 or t
t	tonne	TNE	tne
u	unified atomic mass unit	U	u
UA	French usage, astronomical unit	UA	ua
V	volt	V	v
V_o	volume of ideal gas standard conditions	VO	vo
W	watt	W	w
Wb	weber	WB	wb

FIG. 10. (cont.)

The SI rules for usage of the system are an important factor in the conversion of the computer system. The following rules of metric practice have been adopted by IBM and appear in the IBM publication, *SI Metric Style Manual.*

Punctuation

1. Use a space to separate large numbers into groups of three. Do not use a comma. The comma, where used, is recognized as a decimal point.

   Example: Correct   Incorrect
              600 300.0   600,300.0

2. Use a decimal to express partial units. Avoid using fractions to express partial units.

   Example: Correct   Incorrect
              0.25 kg   ¼ kg

3. Space between the numeric value and the unit symbol.

Example: | Correct | Incorrect |
|---|---|
| 1 mm | 1mm |
| 1 A | 1A |

4. Use of uppercase or lowercase letters as given in the tables on page 6 must be followed to avoid confusion. All abbreviations or symbols are to be used in the singular form. Do not use periods after symbols unless they conclude a sentence.

Example: | Correct | Incorrect |
|---|---|
| mm (millimetres) | mms or mm's |
| ms | m.s. |

Spelling and Capitalization

Refer to the table on pages 5 and 6 for proper spelling, capitalization, and pluralization of SI units.

1. When using the symbols for ohm (Ω), micro (μ), and other characters of the Greek alphabet, it is preferable to use a typed symbol if available. If it is not available, you may spell out the word. If it is necessary to hand-draw symbols, refer to the IBM Corporate Engineering Practice manual, section CEP 0 - 1002. Hand-drawn symbols should be avoided in formal publications. Do not use a lowercase u for micro.

Example: | Correct | Incorrect |
|---|---|
| μs or microsecond | us |
| Ω or ohm | |

2. Avoid capitalization of unit names (except Celsius) unless they start a sentence. Listings should be lowercase.

Example: | Correct | Incorrect |
|---|---|
| The base unit for mass is the kilogram. | The base unit for mass is the Kilogram. |
| The following are the seven SI base units. | The following are the seven SI base units. |
| metre | Metre |
| kilogram | Kilogram |
| second | Second |
| ampere | Ampere |
| kelvin | Kelvin |
| mole | Mole |
| candela | Candela |

3. Unit names are not capitalized (except for Celsius) unless used to start a sentence. Unit abbreviations (symbols) derived from proper names are capitalized.

Usage and Format

1. Use a zero in front of the decimal point when the numerical value is a partial unit.

Example:	Correct	Incorrect
	0.6 mm	.6 mm

2. Avoid using multiple prefixes.

Example:	Correct	Incorrect
	pF	$\mu\mu$F

3. In compound units formed by division, use the solidus (/) or show the division as a negative power with the product dot (raised period). Do not use more than one solidus in a combination.

Example:	Correct	Incorrect
	m/s	$\frac{m}{s}$
	$m \cdot s^{-1}$	$m \div s$
	$W \cdot m^{-2} \cdot K^{-4}$	$W/m^2/K^4$

4. In compound units formed by multiplication, use the product dot (raised period).

Example:	Correct	Incorrect
	m·K (metre kelvin)	mK (millikelvin)
	W·b (watt barn)	Wb (weber)

5. Avoid using prefixes in a denominator.

Example:	Correct	Incorrect
	km/s	m/ms

   Exception: The exception to the rule of avoidance of prefixes in a denominator is the prefix k in kg (kilogram).

6. Prefixes must be combined with units.

7. If dual dimensions are used in text, show the non-SI unit in parenthesis.

Example:	62 mm (2.44 in)
	1 kg (2.205 lb)

8. Symbols for SI units should be printed in upright type, regardless of the type used in the rest of the text. Typestyles such as script or italic are unacceptable.

9. When writing or typing tolerances in text or drawings, it is preferable to place the positive tolerance above the negative tolerance.

   Example: $49.5 \begin{smallmatrix} +0.4 \\ -0.2 \end{smallmatrix}$ mm

   However, in single spaced copy, tolerances should be placed on the same line. For example 49.5 + 0.4 −0.2 mm (positive value first) or 49.5 ± 0.2 mm if both positive and negative tolerances are the same.

   Zero tolerances should be shown.

   Example: $49.5 + 0.4 -0$ or $49.5 \begin{smallmatrix} +0.4 \\ -0 \end{smallmatrix}$ mm

10. Avoid mixing units, symbols, and words.

Example:	Correct	Incorrect
	12.75 m	12 m 750 mm
	metre per second	metre/second or metre/s
	m/s	

11. When showing mathematical operations, use only symbols.

Example:	Correct	Incorrect
	° = rad·π/180	degree = radian x π/180

12. Avoid mixing prefixes within a text or drawing.

Example:	Correct	Incorrect

13. Choose prefixes that give numerical values of 0.1 through 1 000 except when it results in mixing of prefixes in text or drawings.

    Example: 1.2×10^4N can be written 12 kN
    0.003 94 m can be written 3.94 mm
    1 401 Pa can be written 1.401 kPa
    3.1×10^{-8} s can be written 31 ns

Computers

Now that there is an understanding of the metric system, the computer system should be considered next. The computer system consists of hardware and software. Hardware consists of the fixed boxes comprising the computer system: the computer, printers, memory devices, etc. Software consists of the programs which are developed to drive the computers to accept input, process it, and produce an output.

Almost any company today purchases and uses many items which are obtained and/or measured in the metric quantities of grams, liters, etc. Laboratories and some manufacturing areas calculate and measure using metric formulas and instruments. This metric information is prepared, fed through the computer system, processed, and stored and/or produced in some form of output regularly in our present environment.

This chapter will not attempt to give a complete description of computer operation. However, very briefly, the operation is as follows: The preparation of the data is by means of typewriters, typewriter-like keyboard devices, or keypunch machines. The computer accepts the information directly from some devices or indirectly from punched cards, magnetic or punched tapes (numerical machine tapes), or other storage devices. The information is then processed computerwise and is placed into tapes, discs, or other memory devices, displayed upon cathode ray tubes, and punched out on cards or printed onto forms or cards. The use of the metric system may or may not affect all, some, or none of the above.

Numerical Control Graphics

Numerically controlled N/C machining centers, plotters, and graphic displays are operated by information generated by a computer-assisted programming language. The N/C programs are developed and coded by the N/C engineer, keypunched into cards, and sent to the computer to be processed. During the computer processing, a file is addressed to automatically pull out a program to post-process the program for the machine tool this program is to run on. Files containing previously stored data may be called upon to automatically give output previously available only through manual programming. The computer output is available in several forms:

1. Punched paper tape
2. Cards
3. Magnetic tape to support a TP Operation
4. Stored in a library (memory) within the computer for future use

Numerical control machines are designed with inch controls, feeds, speeds, etc.; new equipment is being designed as metric or inch/metric switchable. This means that the inch machines will accept only inch input, metric machines will accept only metric input, and the switchable machines will accept either inch or metric when switched to the inch or metric, respectively, but not vice versa.

All tools and tool data for N/C use are currently stored in memory only in inch measurements. During the changeover, some or all of these tools and information will be desirable to use "as is" until existing stocks are worn out.

Calculations

Existing computer equipment can handle mathematical calculations of some kinds of metric information. Formulas and tables of customary values, sizes, and other constants are programmed and stored for reference and recall under Numerical Control. Much of this material will require varying degrees of conversion.

Historicals

Past programs have been developed to use inch-based information almost exclusively; the exceptions are rare. Thus, many programs must be considered for possible conversion. Existing formats for the inputs and outputs of these programs are usually at their limits, and most changes to them will require considerable justification before being rewritten. This applies to the reports, orders, bills of material, and all other output in existing format for all areas of the company.

Conversion of Computer Operations

Conversion of computer operations will require changes in both hardware and software equipment. In general, the basic hardware will be unaffected. The exceptions are input/output (I/O) devices, which will need different character sets: uppercase, lowercase, and Greek characters. Also, there will be a need for certain mathematical symbols such as the raised minus sign indicating a negative power, the lowered (subscript) numbers or letters, and others. Alternate symbols are not always practical. Most output units only have uppercase characters. Probably no standard I/O device has Greek characters, unless they have been specifically ordered.

The need for the different character sets presents three unique problems:

1. The problem of obtaining I/O devices with the desired character sets is not too serious, as most I/O manufacturers will provide chains, trains, keyboards, fonts or balls, etc., which have the desired characters. There is usually an extra charge for changing an existing set or buying an additional one. Ordering a unit with the required character sets when getting a new computer installation is probably the most desirable and least costly approach.

2. The problem of throughput resulting from the use of different character sets is next in the order of seriousness. A chain or train of characters usually has three identical sets of alpha/numeric characters. Replacing one of the sets, for instance, with lower case alpha and Greek characters would, because of the printer design, slow down the resulting output and consequently reduce the overall throughput of the system. This can be costly.

3. The use of some alternative technique instead of changing character sets would be the next approach. This could be the use of the approved alternate symbols (Fig. 10). These alternates could, in some usage, create undesirable situations. When this occurs, the term should be spelled out, i.e., ohms or micrometers, instead of using the required character symbol, which would be the first alternate for the proper symbol, mm. The problem with this approach is that considerably more space is required. (The space problem will be covered below in more detail.)

The area of software, especially input programs and the output, will be most affected by the change. There are several reasons for this impact:

1. For a long time we will be living in a dual customary/metric world and there will be a need to have both systems of measurement on the same document in already existing columns or fields. This condition creates a need for additional lines of space and/or wider fields to handle the information. Another requirement is to identify the measurement system (metric or customary) of the unit or quantity. Probably the easiest method would be to use a unit of measurement code. A typical process document shown in Fig. 11 illustrates the situation of extra lines for information for dual measurement and also the extra columns for the unit of measure code. (See Fig. 12 for typical unit of measure code.)

    The first line of the operation carries the unit of measure code 32 which indicates that the following quantity is in pounds. Line 100 gives

MACH GRP	SER	TR	C STD	TOOL MMR CLAS	DESCRPTION	R C	TOOL/RAW MAT/ MASTER NO.	DWG	U/M	HRS/100 WT/100	SET UP	A
000	03				RAW MATL	G	04-0110253822		032	9.900		
010					01-253 ALUM	D						
020					44.45 MM (1.750 IN) DIA	D						
030					28.25 MM (1.112 IN) STK	D						
040					PER PC	D						
100					011253 822 42 4.491	D						

OPERATION: 0020   DEPARTMENT: 0050   CDOPT: OVLPP: OFSTT: +000   USE CODE: 00

MACH GRP	SER	TR	C STD	TOOL MMR CLAS	DESCRPTION	R C	TOOL/RAW MAT/ MASTER NO.	DWG	U/M	HRS/100 WT/100	SET UP	A
0173 000	01	L		2	R END OUT-FACE-SPOT-DR-	A						
010					BORE-REAM 12.70 MM DIA	I						
020					(.5000 IN) TO 12.40 MM	I						
030					PLUS 0.0254 MM (.488	I						
040					PLUS .001 IN)-TURN OD	I						
050					TO 41.28 MM PLUS MINUS	I						
060					0.0508 MM (1.625 PLUS	I						
070					MINUS .002 IN)-BRK COR	I						
080					ON HOLE LIGHTLY-FORM	I						

FIG. 11. Dual measurement process document.

01	Piece-Each		51	Pint
02	Pair (2 Pieces)		52	Quart
03	Set		53	Gallon
04	Roll		54	Half Gallon
05	Sheet		55	Imperial Pint
06	C (100 Pieces)		56	Imperial Quart
07	Gross (144 Pieces)		57	Imperial Gallon
08	Ream (500 Sheets)		58	Imperial Half Gallon
09	Bale		59	Fluid Ounce
11	Inch		60	Imperial Fluid Ounce
12	Foot		61	M (1000 Pieces)
13	Yard		62	0.1 Piece
14	Sq. Yard		63	0.01 Piece
15	Bd. Foot		66	Cubic Inch
16	Sq. Inch		68	Dram — Apothecaries
17	Sq. Foot		69	Ounce — Apothecaries
18	Cu. Foot			
19	100 Feet		70	Pound — Apothecaries
21	Meter (m)		81	Box — Package
22	Centimeter (cm)		82	Carton
23	Sq. Meter (m^2)		83	Tube
24	Sq. Decimeter (dm^2)		84	Barrel
25	Sq. Centimeter (cm^2)		85	Drum
26	Millimeter (mm)		86	Tank
			87	Can
28	Sq. Millimeter (mm^2)		88	Carboy
			89	Keg
31	Ounce — Avoirdupois			
32	Pound — Avoirdupois		91	Dozen (12 Pieces)
33	100 Weight		92	Pad
34	Ton — Short		93	Spool
35	Ounce — Troy		94	Ball
36	Pound — Troy		95	Jar
37	Dram — Avoirdupois		96	Bottle
			97	Card
40	Grain		98	Label
41	Gram (g)		99	Lot
42	Kilogram (kg)			
43	Cubic Decimeter (Liter) (dm^3) (l)			
44	Cubic Centimeter (Milliliter) (cm^3) (ml)			
45	Carat			
47	Decagram (dag)			
48	Milligram (mg)			

FIG. 12. Unit of measure code.

the unit of measure code as 42, indicating that the quantity following is in kilograms. Lines 020, 040, and 060 show the additional lines required for the dual process measurements. Notice that the alternate symbol MM is used instead of mm for millimeters to identify the metric dimensions and IN is used to identify the inch dimensions. Not obvious to the reader is a subroutine in the program which will identify lines 000 and 100. When complete conversion to metric only is desired for raw materials, the subroutines will drop line 000 and move line 100 up into position to replace it.

SI Metric in the heading (not shown) signifies that SI standard rules have been applied throughout the document.

These techniques can be used in many commonly used Information Systems Documents such as process documents, bills of materials, etc. It stands to reason, therefore, that if the input was via punched card, extra fields would be necessary for the unit of measure code. Lacking space for the field, an extra card would need to be punched to handle the information. Furthermore, an extra card would be needed for each extra line. The output documents would also require reformatting to handle the information and, of course, modification of the program is required and additional core storage and transactions would become necessary.

2. There are programs in existence which already carry both the metric and customary measurement systems. These present a unique problem in that little, if any, of the old metric usage conforms in the area of SI style. Examples of this are found in the use of correct symbols, punctuation, spelling, capitalization, and also the correct usage and format rules. As before, this results in the need for increased field sizes, additional space, and program changes.

3. Measurement conversions from either system to the other creates another field requirement problem. Conversions from inch to metric create the need for more fields on the left side of the decimal point, i.e., 39.37 inches = 1000 mm.

While it is possible to change this to higher values such as 1 m to reduce field requirements, this may be undesirable because mixing values in drawings and text is a possible source of confusion. Conversions from metric to inch require more fields on the right side, i.e., 0.01 mm = 0.0004 in. This is complicated even more if the rounding rules for greater accuracy are used (in this case 0.01 mm = 0.00039 inches).

4. Numerical Control machines and Graphics design equipment, because of design of internal circuitry, are readily converted by writing a routine for

the post-processor. By adding an identifying code (see below) in the source program for the units in and units out which are desired, the post-processor will make the conversion from the inch or metric input. The output then can be in any of the following forms:

Code 0	inch input	inch output
1	metric input	inch output
2	inch input	metric output
3	metric input	metric output

If, for instance, a job with an inch-based program is running on an inch machine but the load is such that the job must also be run on an identical metric machine, the program can be changed by the post-processor to metric output by merely changing the identifying code in the source program.

In brief, then, the desired units code is placed in the source program, a permanent routine written for the post-processor reads the identifying code and processes the input units into the desired output units without any hardware changes.

5. Computer mathematical calculations will have varying degrees of impact. Many mathematics programs will have very little need for change simply because the new measurement language will not change basic calculations. Where it is necessary, the modification can easily be made.

Programs involving built-in tables of customary values, sizes, or other constants will not be readily converted to exact metric values. In some cases, the exact value is really not necessary and consideration should be given to inputting new tables of exact metric values rather than trying to develop complex conversion programs. For example, except for historical comparisons, square feet or square yards would not be convered to the odd 0.836 square meter size. Rather, the full square meter with the cost/ square meter would be the necessary metric value for future calculations. Also, instead of converting 0.017-inch wire to 0.4318 mm, the designer would select a preferred size of 0.4 mm or 0.5 mm.

Routines can be written into existing programs to handle much of the conversion, but many programs will need considerable modification and/or rewriting.

6. Considerable thought has been given to the need to convert historical data to metric measurement ostensibly for comparison purposes. When this is necessary, it will probably be done manually or with a simple desk calculator on a one-time basis.

Other Problems

Field Size

Conversion to metric will result in field sizes different from those now in use. When fields are larger the following conditions must be addressed:

1. Data files: require larger records, changed maintenance transactions, additional file space, expanded file reports/listings
2. Transactions: very probably an additional number of transactions due to larger field sizes
3. Reports: all must be changed to reflect metric notation and field sizes

Conversions

This will only be a small problem if they are uniformly expressed in metric units. However, they will be more of a problem if the product has components in customary and metric units of measure. This affects such documents as bills of material, process routings, and inventory listings.

Rounding

Rounding error resulting from conversion to metric/customary units must be carefully analyzed against correct rounding rules to maintain required accuracy. In some cases, conversion tables and program rules are necessary to insure that conversions are made so that the Information System is compatible with the capabilities of Numerical Controlled machines.

Parts Interchangeability/Usage

If a part is to be used within both the metric and customary product, the Information System must support a dual description of this part or have conversion capability. Decisions such as these must be based on considerations such as least cost, reduced lead time, etc.

Conclusion

Even though existing programs may carry both metric and customary measurements, conversion of computer operations should not be dismissed as "no problem." As this chapter has shown, there are some very real problems. These problems can be resolved, but only if the functional users bring them to the attention of the systems people. Recognition of the problems is a big step toward solving them.

Chapter 7

METRICATION TRAINING

DAVE L. VENTON

Technical Education and Training Department
Xerox Corporation
Rochester, New York

When a company decides to convert to the metric system, one of the first potential problem areas to surface is the training of personnel. This is as it should be, for proper training can be the key to a successful transition, minimizing costs and reducing or eliminating other problems.

Successful metric training will depend to a large extent upon the support given by management. Management commitment is required early in metric transition to provide direction and resources to the training department. Nothing is more frustrating for a training manager than an identified need without the resources to fill that need.

A sure sign of commitment by management is the appointment of a metric coordinator and the forming of a metrication committee. The committee, chaired by the metric coordinator, should set policy, establish conversion schedules, and provide guidance to the training manager. Ideally, the training manager should be a member of the metrication committee. In this way, he or she will be abreast of the latest decisions that may have impact upon the training department.

Once a metric conversion schedule has been established (what products? when?), the training manager can establish a starting date for metric training. With this in mind, a metric awareness program should be started to inform the employees of the impending change. This awareness phase, if properly administered, can do a lot toward reducing the resistance to change that is normally encountered during the transition to metric.

The training department must now define the learner population and identify the actual metric training needs. A properly prepared needs analysis can do much to reduce training costs and the time required to complete the program.

Once the learner population has been defined and the metric training needs determined, a decision must be made as to the method(s) of instruction. Small groups of people requiring a minimum of training may dictate classroom lectures, while large groups requiring in-depth metric training may dictate a more expensive and sophisticated approach, such as programmed instruction or video tapes. On-the-job training may be the logical approach for other groups. At this point, an estimate should be made of the time and funding required to complete the training. This should be agreed to by the metric committee prior to actual program development.

Once approval has been received, development of the programs should begin. Different programs will be required for different segments of the population. By careful structuring and a little manipulation (using the needs analysis data), certain portions of the training program can be reduced or expanded as needed so that a basic program will fill the needs of a relatively broad section of the learner population. For example, almost everyone should be exposed to at least an introductory program covering the most commonly used units of the metric system. This unit of training could then be made the first of a series of training modules covering other metric units in more detail. A student would take only those modules identified by the needs analysis data. The number of training modules required and their instructions/content can be modified to fit the exact learner population to ensure that each student receives the proper level of instruction.

It is imperative that each student be trained *just prior* to his or her actual on-the-job need. Training too soon is almost as bad as training too late. This sounds simple, but, in fact, can be very difficult. If your company, like most companies, is planning to convert to metric only on new product programs, then you have a real training problem. Once everyone on your first metric product is trained, you will need to train new persons assigned to that program as they receive assignments. In addition, as other new products are started, the people on these new programs must also receive timely training in metric units. In addition, engineering and drafting personnel must be trained during the development phase of a new product, while manufacturing and sales personnel will require their training just prior to production or introduction of that same product. The logistics are almost staggering. As you can see, metric training is not a one-time thing, but a continual effort that could conceivably last for years. This must be taken into consideration when determining the method of presentation. The costs of a canned program can be prorated over several years, actually saving your company time and money compared to continuous classroom lectures.

The information contained herein is designed to help the training manager overcome many of the problems associated with conversion to the metric system. It is a systematic approach to metric training and can be adapted for use in most industry in America.

Gaining Management Commitment

It is imperative that the management of your company be committed to the transition from domestic to metric measurement. Without this commitment, all attempts to develop training programs and/or train personnel in the metric system of measurement will be frustrated. The training manager must be assured of adequate fundings, training personnel, and other resources if he is to do the job required.

A company that is definitely committed to metric transition will normally have: (1) announced formally that metric conversion will take place; (2) formed a metrication council or committee to provide direction during the changeover; and (3) appointed a metric coordinator to chair the committee and function as the company's metric spokesman.

The metrication committee should consist of representatives from engineering, manufacturing, quality control, standards, training, and marketing. In other words, a cross section of corporate functions should be represented if the committee is to be effective.

The chairman of the committee (metric coordinator) is management's representative; he should have enough authority in the company to be effective.

If such a commitment to metrication is not evident, then it might be advisable for the training manager to draft a proposal for metric training. The proposal should be directed at top management, and should be very specific in terms of projected costs and impact. The proposal could be dual in nature, comparing costs, time, etc., between a coordinated effort and a hit-and-miss approach. The proposal must be logical and direct, and must sell the concept of total commitment to management.

Establish a Conversion Schedule

A schedule for metric conversion must be established by the metrication committee and approved by upper management before the training manager can take action. He must know what new product or products will be affected, and when. Obviously, the product must still be in the early stages of concept or development if it is to be designed using metric units.

A close working relationship must be established between the training manager and the new product engineering manager. The metric training start date should be coordinated with the product development schedule so that affected personnel can be trained just prior to their actual need to know. The metric training schedule should first include such technical personnel as engineers, draftsmen, technicians and modelmakers; training of manufacturing and quality control personnel would follow. The last to need training in metric will probably be marketing, sales and/or service personnel. At this point, your schedule will be tentative only, but it will be a base upon which you can establish your manpower requirements and projected costs. The metric training schedule should be concurred upon by the metrication committee.

Begin a Metric Awareness Program

Once the commitment to "go metric" has been made and a tentative training schedule established, employees should be made aware of the impending change. An employee "metric awareness" program should be started. This will do much to reduce the normal resistance-to-change you can expect to encounter during the metric transition.

This metric awareness program need not be too structured and should take a minimum of time and effort. The idea is to inform your employees of what the company is doing and to arouse their curiosity concerning the change to metric. Articles concerning the metric changeover should be written and published in your company newspaper; metric wall charts can be posted conspicuously near bulletin boards; and metric movies can be rented and shown during lunch breaks; paper cups with metric slogans can be put in the coffee machines; displays of metric scales, instruments, etc., can be put in the lobby; and so forth. Use your imagination; at this point, metrication is a sales, rather than a training, job.

Define Your Learner Population

Before you can start metric training or training development, you must determine your learner population. Obviously, different jobs will have

different metric training requirements. Before establishing the actual requirements, you should break down your potential students into some kind of
categories, usually by job category or job title. A typical breakdown might
be as follows:

Nontechnical clerical
Technical clerical
Draftsmen
Technicians
Engineers
Technical management
Middle management
Foreman/Supervisors
Machinists
Tool and die makers
Quality Control personnel
Sales and Service personnel

This list can be modified to fit your particular company or area of responsibility. In any event, such a breakdown is required if you expect to perform
the next step, which is needs analysis.

Determine Metric Training Needs

A properly prepared needs analysis is a necessity if you are going to conduct
a successful metric training program. The object, of course, is to train only
those people who require metric training, in only what they really need to
know to perform their job, and to accomplish the training in a timely manner.

By using questionnaires directed at specific job categories, you can
establish the basis for a task analysis for each job. The objective here is to
determine exactly what a person does in the daily performance of his or her
job, how frequently they perform a task, how critical the task is to the
success of the total job, and the projected impact on the company if the
function were performed incorrectly.

After eliminating all tasks that are not affected by metric practices or
measurement, you will have established the core of metric knowledge
required for each job. Using this information, selected group interviews can
be conducted to further refine your data.

This needs analysis can now be plotted on a matrix of metric units, terms,
symbols, etc. to determine commonality of training requirements across

	Nontech Cler	Tech Cler	Draftsmen	Technicians	Engineers	Tech Mgmt	Mid Mgmt	Foremen/Supv	Machinists	T and D Makers	Qual Ctrl	Sales/Serv
Seven Base Units—Overview		•	•	•	•	•	•				•	•
Meter—Overview	•	•	•	•	•	•	•	•	•	•	•	•
Meter—In Depth			•	•	•	•		•	•	•	•	
Kilogram-Overview	•	•	•	•	•	•	•	•	•	•	•	•
Kilogram—In Depth			•	•	•	•					•	
Second—Overview	•	•	•	•	•	•	•	•	•	•	•	•
Ampere—Overview		•	•	•	•	•	•	•			•	•
Ampere—In Depth			•	•	•	•						
Kelvin—Overview			•	•	•	•	•				•	
Celsius—Overview	•	•	•	•	•	•	•	•	•	•	•	•
Celsius—In Depth			•	•	•	•					•	
Mole—Overview					•	•						
Candela—Overview					•	•						
Candela—In Depth					•							
Liter—Overview	•	•	•	•	•	•	•	•	•	•	•	•
Liter—In Depth			•	•	•	•					•	
Radian—Overview			•	•	•	•	•				•	
Radian—In Depth					•							
Steradian—Overview			•	•	•	•	•				•	
Steradian—In Depth					•							
Multiples and Submultiples—Overview	•	•	•	•	•	•	•	•	•	•	•	•
Common Usage—Multiples and Submultiples—Overview	•	•	•	•	•	•	•	•	•	•	•	•
Common Usage—Multiples and Submultiples—In Depth		•	•	•	•	•	•	•	•	•	•	
Common Derived Units—Overview		•	•	•	•	•	•	•			•	•
Common Derived Units—In Depth		•	•	•	•	•	•	•			•	
Derived Units—Overview			•	•	•	•					•	
Derived Units—In Depth					•	•						
Symbols—Overview		•	•	•	•	•	•	•			•	•
Symbols—In Depth		•	•	•	•	•					•	
Micrometer					•				•	•	•	•
Vernier Caliper					•				•	•	•	•
Dial Indicators									•	•	•	•
Reading Engineering Drawings				•	•	•			•	•	•	•

FIG. 13. Needs analysis matrix of metric units.

different jobs. An example of a typical matrix is shown in Fig. 13. The metric units and other types of units for which training may be needed are listed on the left. The job categories are listed across the top. A check mark is placed in the appropriate blocks according to the needs analysis data for each job. This will establish a fairly comprehensive picture of metric training needs for your organization.

Using the information checked in the chart (Fig. 13), you will see that, for this example, there are seven items that are needed in every job category. This would indicate that a metric training program consisting of these seven topics would be applicable to everyone. This then, could be the first module of a multimodule training program. Eventually, additional modules would be added to cover the needs of everyone.

Establish Training Method

Having defined your learner population and their specific metric training requirements, you are ready to select a method or methods of actual training. If you have a fairly large population to train, or if the training will be conducted over an extended period of time, it may be advisable to use programmed instruction techniques or videotape. Relatively small numbers of students might dictate classroom training using standard teaching methods. Other segments of your learner population might better receive on-the-job training, particularly where only a limited amount of knowledge is required.

You may find that there are programs already developed and on the market that will fill some or all of your metric training needs. For example, there are several commercially available programs concerning the use of a metric micrometer and/or vernier caliper. Whenever possible, review such programs for possible use in your company. Even if they are not appropriate, you will gain usable knowledge in the subject matter than can be applied to your own programs.

Whatever your final decision, remember to keep the cost per student hour as low as possible. Review all methods carefully before making your final decision.

Estimate Time and Costs

You are now ready to estimate how long it will take to train your employees, and how much it is going to cost. You should know who you are going to train (and how many), what your are going to teach, and what method or methods of training are going to be employed.

Estimate how long it will take to develop the metric training programs. How many hours of instruction will be required for each job category? How many people over what period of time will have to be trained? This should then be weighed against the training schedule previously developed.

In all likelihood, you will either have to modify the schedule or reestimate your development and training time. You may have to increase your training staff to meet the scheduled requirements. If you cannot change the schedule and additional headcount is not available, you may have to change your training method or decrease the amount of metric training each employee will receive.

Once you have managed to get the schedule and your development and training time to match, you should estimate the costs for the program. This estimate, along with the proposed program and schedule, should be approved by the metrication committee before further action is taken.

Write the Program

When approval is received from the metrication committee, you can start the actual writing of your training program.

Break up your subject matter into logical pieces, establish an outline for each segment, and tie them all together in proper sequence. Different modules can be given to different program writers to shorten the actual development time. But be careful to maintain program continuity. The writers must work closely together if you are to avoid redundancies and end up with a well-structured, meaningful training program. ANMC's *Metric Education Guide For Employee Training* is a useful reference.

Conduct Training

The important thing here is to train a person just prior to the actual need-to-know. A "shot gun" approach will generally be a waste of time for you and for the learner. This timely training is not as easy as it appears on the surface.

Suppose your company has decided to develop and market a new metric product. Only the people directly associated with that new product will require training in SI metric. Initially, only a handful of engineers, draftsmen, and technicians assigned to the new product may require training. Later, however, as the product program grows, you may have additional personnel assigned to the program that will require immediate training; you will be expected to drop everything and train them. As the new product nears the end of development, you will be required to train modelmakers and tool and die makers. When you start into production, you must train machinists, foremen, inspectors, etc., and just prior to distribution, you must train the sales and service personnel.

During this entire program, you must be prepared to train each new person as he or she arrives on the scene, through new hires, transfers, etc.

This same sequence of events will take place each time a new metric product is introduced, until some time in the future when all programs are metric. This obviously could take many years.

All of these must be considered when determining your method of metric training. Short, modular programs, using videotape or programmed instruction, will do much to reduce what could be an overwhelming workload.

The thing to remember is that metric training will be a continuous effort for several years. Additionally, each affected employee must be trained in a timely manner during the transition phase. This is a monumental task, even in a small company. It can be accomplished, however, by careful planning, close liaison with the metrication committee, and properly prepared modular training programs.

Review

This approach to metric training will only be successful if it is systematically followed and has full management support. In review, the following steps should guarantee you some measure of success during the critical period of conversion to the metric system of measurement.

1. Gain management commitment
2. Establish a conversion schedule
3. Begin an awareness program
4. Define your learner population
5. Determine training needs
6. Determine training method
7. Estimate time and costs
8. Gain metrication committee approval
9. Write the training program
10. Conduct training in a timely manner

Chapter 8

METRICATION COST MANAGEMENT

JOHN T. BENEDICT

Engineering Office
Chrysler Corporation
Detroit, Michigan

Weights and measures would probably be a prime candidate for selection as the subject most people take for granted. Yet, recognized or not, the basic system of measurement is a mainstay of our modern economy and a key underlying factor in civilization as we know it today.

It has been estimated that more than 20 billion measurements are made daily in the United States—in industry, science, commerce, government, education, and everyday living. Who makes all these measurements? Who uses them? The answer, of course, is that we all do. In a highly developed industrialized society, the customary or nationally accepted measurement system and its myriad applications are vitrually all-pervasive.

Although the metric system has been legal in the U.S. for more than 100 years, the Imperial inch-pound system has been in customary use as the predominant system since the country's founding. Now however, after two centuries of debate in government and industry circles, the inevitable changeover to metric measurement is occurring, as metrication gathers momentum rapidly throughout the United States and the national commitment to proceed is clearly established.

At this point, we recognize that the opportunity to improve something as basic as our measurement system and the associated engineering standards is exceedingly rare. In fact, for most people, it does not happen even once in a lifetime. Thus, without exaggeration, it might be said that metrication, and the growth of international cooperation and interdependence that now are upon us, are the greatest changes we are likely to see in our own lifetime, and, together, they offer the best opportunity to make a fresh start since the industrial revolution.

These are grand thoughts, expressed in high-sounding words. Let us proceed to direct the cool light of reality on these lofty ideals. From a purely business standpoint, the prospect of entering into metric transition should be viewed as an investment. The attitude toward various proposed plans and methods for metrication within a company's products and operations should be guided by one overriding consideration: is the investment justified economically? Accepting the necessity for metrication, a company's top management must look upon alternate approaches and judge them mainly on business merit. What will be the advantages and disadvantages, on both a short-term and long-term basis, to our company? What investment is required? What can we expect to gain as a measurable return on this investment? What problems will we encounter?

An objective view is needed. Extraordinary effort should be exerted toward making a truly informed decision. In the final analysis, the foregoing basic questions dealing with investment and returns must be answered to reach a rational solution for the metrication issue that increasingly confronts U.S. businessmen and demands their attention.

From a management viewpoint, metric conversion is a major, long-term project whose impact will be greatest in engineering and manufacturing areas; but which, to some degree, will affect every department in the company. Facing a change of this magnitude, scope, and duration, a businessman knows it must be managed effectively, and that the management objective should be to follow the "least-cost path" while getting maximum value from the change.

This brings to the fore the concept of metrication cost management, which is the topic to be treated here. At the outset, it would be useful to establish definitions for two of the terms: *Metrication* is a broad, rather loose term that can be applied to the overall process of increasing metric measurement usage and changeover to predominant use of the metric system. *Metrication cost* is not the entire cost of a metric product or facilities program, but just that portion (or extra cost) ascribable to metrication.

The Metrication Launch Window

Borrowing an aerospace expression, "launch window," and applying it to metrication, one could say that each company should determine its own individual "metrication launch window"—when all major factors combine in establishing the most propitious time for cost-effective start and follow through on metrication. Main inputs for determining your own particular "window" are:

1. The U.S. national metric trend
2. Your own industry trend
3. New product cycle (with emphasis on predicting the production life cycle and last year a new product will be carried in manufacturing)
4. International product involvement

From consideration of these factors can come perception of your window for metrication launch, a judgment of when to give the "go" signal for implementation and basic understanding about the best pace and duration of your metrication program. These are critical factors, because launch timing, program pace, and duration have a crucial effect on cost.

Philosophy and Rationale

In essence, the metrication cost management philosophy developed from experience of many organizations is: follow the rule of reason and balance strong long-term planning with equally strong, realistic current cost control. In practical terms, this means: plan for metrication, but do not provide either special management or a separate budgetary line item for it.

The change to metric is by no means painless or free of cost. Badly handled, the costs—in terms of confusion and wasted effort, mistakes and botched plans, part variety proliferation, "bi-unital" operations, dual inventories, etc.—could be substantial. These cost factors are quite apart from changing or replacing measurement equipment and production machines.

To make an even more fundamental point: if company conversion to SI metric involves an allocable and auditable incremental outlay of major proportions, there may be alternative investments deserving of a higher priority.

In line with normal good business practice, managers want to know the projected cost of programs they are considering. This is understandable. Usually such information is needed and used to control, justify, recover, or offset costs.

Metrication cost, however, is not susceptible to management and control by techniques that entail quantifying all elements of cost. The costs of metrication are much discussed, widely debated, and quite controversial. Many of the costs are intangible (slight drop in productivity, slight decline in proficiency due to thinking and working in a new system, nuisance factor, inconvenience, duplication and errors while operating with two measurement systems), and virtually impossible to measure accurately.

Experience has shown that precise, realistic cost estimates (before, during, or after) are impossible to determine and, further, that a strong, thorough, and sustained effort to forecast and measure costs is counterproductive.

Metrication cost appears to be a direct function of management attitude. If you direct an organization to convert over a period of years, using common sense, and give them no extra funds for metrication, the costs are unmeasurable. However, if you ask, "How much extra money will you need to go metric?", you'll very likely receive estimates that add up to some sizable sums.

When we have learned what there is to know about metrication cost, and when discussion has quieted down, we should keep in mind the fundamental point: metrication must be accomplished in the normal course of business; there simply is no "separate pot of money" for metrication.

Turning now to a rationale for metrication: by whatever combination of internal initiative and reaction to external events a company arrives at a metric commitment, a positive attitude should guide its ensuing actions. The positive approach is to view metrication as an opportunity for change with attendant benefits that will offset cost of the actual conversion process as quickly as possible. An alert, somewhat aggressive spirit can identify and attain benefits that will elude the passive and foot-draggers. Whether one enters metrication by choice or by necessity, it might be said that "the inevitable" changeover to metric has begun. A positive acceptance can aid in getting best overall value and economic gain from the change.

Cost Categories

The assessment and allocation of costs is probably one of the least understood aspects of metrication. A thoughtful, item-by-item consideration of the following list of metrication cost elements reveals why this is true:

1. Employee training
2. Employee personal tools
3. External standards activity
4. Internal standards activity
5. Measuring instruments
6. Inspection equipment

7. Machinery and equipment modification
8. New machine tools, metric or dual capability
9. Metric supplies and materials
10. Early obsolescence of machine tools and products
11. Revision of computer software and computer-generated reports
12. Decreased productivity while working in two systems
13. Errors
14. Part variety proliferation
15. Warehousing, inventories, and distribution
16. Technical reference and publications revision
17. Man-hours spent on various aspects of metrication

It is evident that metrication cost is not amenable to management and control by techniques that assign "numbers" to all elements of cost and treat them as separate line items for accounting purposes. Although the above list is not all-inclusive, it does convey a sense of the pervasiveness of metrication. It also helps one perceive the comprehensiveness, precision, and intricacy of an accounting system that would isolate and record tangible and intangible cost factors as widespread as those listed above. Keep in mind that, in most areas, metrication costs are of small percentage magnitude. Figure 14 gives the results of a survey on allocating metric costs.

The impact of metrication will be felt on all facets of a company's business and operations: engineering, purchasing, manufacturing, sales, finance, management—all will be affected. Recognizing this, a systematic way to identify elements of metrication cost impact is to use a checklist such as the following list of main tasks:

1. Feasibility examination and preliminary planning
     Adopt metrication policy
     Appoint metrication director
     Form metrication committee
     Prepare general plan
2. Detailed planning and control
     Decide how to carry out metrication
     Assess potential impact in various areas
     Determine how costs will be handled
3. Product design engineering
     Opportunity for variety reduction
     International standardizing
     Dimensions and tolerances
     Engineering drawings

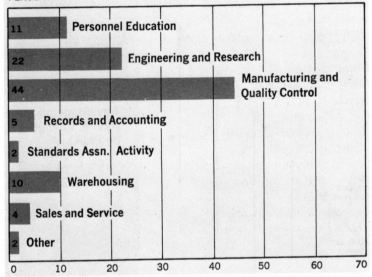

Manufacturing Industry Survey:
Allocation of Estimated Costs of Going Metric

PERCENT

FIG. 14. Manufacturing industry survey: allocation of estimated costs of going metric.

    Design calculations
    Standard parts
    Test and development
4. Production
    Tools, facilities, machinery, equipment
    Factory service, building maintenance
    Storage and handling
5. Purchasing
    Planning the change
    Supplier contacts
    Costing
6. Storage, handling, and distribution
    Stocking inch and metric sizes
    Rationalization—unit sizes
    Design of pallets and packages

7. Computer and data processing
   Assess impact of metric change
   Revise programs
   Standardize with vendors and customers
   Use international standards where possible
8. Sales and Marketing
   Identify measurement-sensitive items
   Pricing
   Distribution
   Advertising
   Service
   Product liability
9. Accounting and administration
   Capital requirements
   Printed matter in SI metric units
   Office supplies; paper sizes
10. Employee training
   Trainees, teachers, materials
   Training requirements and schedule
   Method
11. Literature
   Technical publications
   Promotional literature
   Accuracy

When analyzing each of the functional tasks, costs of metric conversion generally can be considered as:

1. One-time capital costs
   purchase metric measuring instruments
   replace or modify existing machinery, equipment, and tools
   dual inch and metric stock holdings
2. One-time operating costs
   personnel training
   loss of efficiency and wastage due to lack of familiarity with the new system
   replacement of trade literature, stationery, etc.
   replacement of textbooks, handbooks, notes, etc.
   conversion or replacement of specific engineering drawings, etc.
   accelerated obsolescence of equipment and designs
   proliferation of part variety during transition period
   modification of computer programs

Implementing Metric Cost Management

The cost of estimating metrication's potential overall cost prior to implementing a conversion program can be substantial. For a large corporation comprised of numerous operating divisions and many hundreds of departments, a comprehensive, in-depth metric study could cost $100 000 or more in manhours consumed. Experience shows that for most companies, this approach is not justified.

Once a metrication program is underway, the cost of maintaining a mechanism for complete, detailed cost accounting is unwarranted. Further, it tends to be self-defeating, because a "metrication budget," if established, acts as a magnet in attracting costs that otherwise might be covered in regular operating budgets or reduced by a more resourceful approach to a particular problem.

Instead of installing a complete system for tracking metrication costs, some firms feel they control the costs mainly by insisting that there be no separate funding for metrication. Metrication is treated as an integral portion of the program cost for a particular product. Each operating unit is expected to carry out its responsibilities within budget and through local ordering of priorities, etc. Obviously, this philosophy cannot be applied rigidly. There must be exceptions where discrete metrication costs are visible; sometimes they are rather substantial.

Also, it should be noted that some companies do partially record metrication costs and report them as such within their systems for overall product program cost monitoring. In some instances, management desires to know some portion of metrication costs for tax reasons, internal analysis, and statistical reporting, and for possible use in product pricing.

In setting up for effective metrication management (while avoiding the wasteful excesses of overmanagement) the financial aspect is an especially challenging area for consideration. Obviously, the role of financial planning in the conversion process is of great importance. In considering the various different approaches that might be taken, some cost estimates are needed so that the projected outcome of alternate plans can be compared. In many instances, proper consideration of cost implications is assured by placing a high-ranking financial executive on the corporate metrication committee.

In most manufacturing concerns, thus far, the start of the familiar product-driven metrication activity has coincided with the beginning of a new product program or the receipt of metric product orders from major customers. In any case, the setting of the most economic pace for transition is influenced by the normal cycles of facilities and for capital equipment replacement; hence, financial information pertaining to these matters is needed. For optimum results and genuinely cost-effective handling of such matters the application of competent, professional cost accounting expertise is strongly advised.

Proper timing and an orderly changeover are crucial factors in determining cost. The approximate natural order of the time required for any major change is "one generation" of the subject of the change. It is unrealistic to expect a smooth change to be completed in a shorter time, without some degree of compulsion and added cost.

In product design terms, this means we go forward into metric transition with new design programs. Existing inch items are to be phased-out in a normal obsolescence cycle. For each company, there is a certain "natural" rate of metrication: in the context of overall U.S. metrication and its own overseas involvement. Any influence that tends to "force" the transition and accelerate it to a faster rate inevitably will add to the cost.

In light of today's economic reality, it is evident that metrication must pay its own way through early returns for effort applied. The proper timing policy will set the pace of metrication at a level that assures progress at reasonable cost. There is no real basis for establishing the timetable or scheduled endpoint at any particular arbitrary date, such as a point 5, 10, or 15 years hence.

In assessing costs, care should be taken to cover all valid items, including some that are not obvious. Also, it should be recognized that the costs are more likely to be noted than the benefits. Whenever feasible, the provisions for isolating and estimating costs and benefits should move in parallel. In using cost/benefit information as input to determine the optimum transition pace, considerable flexibility can be allowed. The metrication schedule can be shaped by considerations in the areas of resource planning, product planning, marketing, and other key factors.

Financial implications are usually included in an overall report or proposal covering the metrication plan. Subsequently, as metrication progresses, sufficient cost information may be collected and examined to enable judgment on performance against the plan. Again, with suitable caution and constraint against unwise attempts to isolate all costs, it is, nevertheless, desirable to maintain a checking method on actual progress in regard to major items in the financial aspect of the program.

For a large manufacturing company, one of the pervasive, costly items in a complete, long-term metrication program—and one of the most disruptive, prolonged, and difficult to quantify—is the decline in operational efficiency prevalent throughout the entire engineering, manufacturing, purchasing, sales, and management functions. Within a given organization, some people will require a long transition period to regain in metric units the same facility and proficiency they now have with inch-pound units.

Also, with regard to hardware, one of the main problems arises from the long life of capital equipment. Machines may have a useful life of 20-30 years or more. This point again brings the familiar metrication timing dilemma into focus. On the one hand, we know that the most difficult and costly phase of metrication is the time period when operating on two systems; this tells us to make the transition timetable as short as possible. On the other hand, for many businesses, the extended time-span for a complete product design regeneration, coupled with the even lengthier capital equipment replacement cycle may dictate a prolonged transition schedule for the so-called "least-cost path" from start to finish of metric conversion.

No two situations are exactly alike. Each company and organization will have its own distinctive "cost profile" for metrication. Metrication costs will rise to a peak, then reach the tipover point after which, instead of managing the phase-in of metric, we are managing the phase-out of inch items. Probably, at some stage five to ten years after the start of metrication, our operations will be so enmeshed in dual systems and bi-unital items that what many now view as costs will then be seen as savings! Clearly, metrication cost management will continue to be more of an art than a science.

Metrication Cost Estimates

A sampling of cost estimates and some related material are provided in Part IV. Included are selected items from the United States and England, plus a cost profile from the U.S. Department of Defense Metric Study, and a brief extract from the U.S. Metric Study Report on Manufacturing Industry.

In Part IV (Ref. 21 in Chap. 1), a reference is given to one of the more meaningful company metric cost study reports that have been made public. In this instance, a major aerospace firm used two military product programs as "study models." Costs incurred through the design, development, and flight test or pilot production phases were estimated. "Cost factors" were identified where possible; "cost factors" assign a unit value to each task, thus permitting applicability to any program regardless of type or size. For example, the addition of a digital readout capability to a knee and column milling machine is estimated at $6000 per machine (cost factor). This cost is the same on a spacecraft, aircraft, or missile program; the number of machines used depends on the program size. Conversion charts were estimated to "cost" 1½ hour per engineering drawing; general productivity loss, 2% for 4 weeks; training, 4 hours per employee. Similarly, metric impact on numerous functions, documents, and equipment items was assessed and assigned a basic "cost factor." Results of the study indicate that:

1. Most costs are nonrecurring
2. Not all costs incurred can be considered costs of going metric; many are capital expenditures
3. The timing of metric requirements substantially impacts metric costs.

Metrication costs may be estimated or measured against a virtually endless variety of parameters. These range from dollars per employee to percent increase in cost of purchased components and materials; from percent of value added to percent of annual sales over the transition period. Whatever the base and whatever the method, the numbers vary considerably. But generally, one may reasonably conclude that there is some added cost and that, in a given situation, for a particular aspect of metrication, the order of magnitude may range from a fraction of 1% to as much as 10-12% increase. And, to further confound any serious attempt to obtain definitive cost data, the most universal "message" from the voices of experience is that "metrication cost is less than expected."

Such material has been characterized as a bewildering "phantasmagoria of estimates." And admittedly, metrication cost data is typically inconclusive and not definitive. Perhaps the main value from an intensive effort to learn about the cost aspect of metrication is the thorough knowledge and understanding gained of the entire metrication process, along with insights leading to a realization that, although exact measurement and rigorous control of metrication cost should not be attempted, the precepts of sound management are applicable. It becomes evident that a special kind of "good management" may confidently be expected to bring about reasonable success in minimizing metrication's "extra cost" without ever knowing the exact dollar amount.

Opportunities and Benefits

What metrication means to industry is problems, solutions, difficulties, and benefits. Only by concentrating on the benefits can one find the incentive to confront the problems and difficulties. Benefits will not come automatically; they must be pinpointed and pursued aggressively. Good business practice dictates that the position a company takes on the metric question should be based on long-range commercial and economic considerations. As the United

States voluntary metric trend gains momentum with added stimulus from federal legislation, individual company managements are adopting metrication policy commitments. Typically, the policies call for a gradual transition to SI metric measurement units, via a program and schedule paced by the normal cycles of product design and production equipment obsolescence.

At Chrysler Corporation, for example, we are in the early stages of a gradual, long-term transition to the SI metric measurement system. We are keeping in close touch with metric developments in the U.S., and laying the groundwork for an orderly changeover to metric in line with the trend underway in the motor vehicle industry.

Chrysler's program is product-oriented, and emphasis is being placed on selective adoption of international standards for corporate use worldwide. The rate of increase in Chrysler's metric unit usage is being determined by: timing of product programs, supplier capability, the general trend in this country, and the degree of international involvement in a given product program. We expect to derive long-term advantages from use of a common measurement language and uniform standards for engineering, manufacturing, and marketing operations throughout the world.

As the program progresses, metrication becomes a part of every manager's responsibility. It is regarded as a regular part of his job. It is to be accomplished as an added function, performed within the existing organizational structure, and through normal channels and working relationships.

Generalizing again: metrication can be seen as a great challenge and a long-term opportunity. The challenge lies in attempting to obtain maximum value from such far-reaching change. The opportunity comes from the once in a lifetime occasion for "starting with a clean sheet of paper," and attempting to simplify operations by substantially reducing the variety of parts, etc.

Certainly, it is evident that complete metric conversion will be a long and difficult process. It will cause some added cost. But all-in-all, we ultimately can look forward to long-term benefits that will more than justify the cost and effort.

Aside from the much-discussed opportunity for variety reduction, and the equally apparent benefit in standardization of products and components within a given company's world operations, another potential gain to be made is that metrication can be an occasion for introducing long-deferred changes in wide-ranging areas, including product, systems, and hardware. This comes about because the changeover involves far more than a change of dimensions. Adoption of the metric system provides a stimulus for critical examination of existing designs and methods. It can lead to product deproliferation, component standardization, and the streamlining and updating of manufacturing processes and standards.

Conclusion: A Benefit-Oriented, Least-Cost Metrication Strategy

To yield benefits, metrication must be planned and carried out in an orderly manner. For management purposes, metrication may be viewed as a major project of 10-15 years duration and of such proportions that, to some degree, every department in a company will be affected. Development of a full understanding of what is entailed is a first step toward devising a least-cost strategy for metrication. In preparing a program best suited to particular needs and objectives, metrication should be treated as an investment that is keyed to a well-defined set of objectives or benefits. These can be direct benefits obtained from the metric system's simplicity, and from use of a common measurement language in the engineering, manufacturing, and marketing operations of a multinational company. Indirect benefits or "spin-offs" are the things that can be accomplished while undergoing metrication. They consist of a multitude of opportunities for variety reduction in many phases of product design and production, plus significant systems simplification. International standards and product integration on a world basis are essential factors in transforming metrication's benefits from "potential" to reality.

The accepted general management strategy is to embody metric management into the existing organizational structure. The objective is to undergo a realistic and practical change. The advantage of this approach is direct management control, which enables cost-effective programming. In the metrication context, references to cost control do not imply total accounting of all costs (or benefits); this is not practical or realistic. Instead, the key is specific control of expenditures and potential costs.

Metrication implementation strategy is based on normal product flow as the critical path. Cost accounting operates under normal controls and procedures. The program is timed by movement of designs through the stages of planning, management decision, engineering, test, production release, manufacture, quality control, marketing, and service. By using product flow as the activity control, tasks and timing are clearly identifiable in the usual manner. A parallel strategy encompasses other major functions and develops across-the-board guidelines and programs in such areas as purchasing, standards, training, and communication.

Critical importance is assigned to joint planning, good communication, and active coordination among company and industry groups whose metric changeover programs are interdependent. The key to control of costs in any firm engaged in metric conversion is the development of a master plan that is dovetailed into its industry sector program and transition timetable.

Costs of the changeover are evidenced in efforts to overcome inertia and resistance to change, accumulation of metric data, timing mismatch, mixed design interfaces, learning factors, and, in general, a host of matters attendant to operating with two measurement systems. One system is superior to two systems, and there is consensus that the benefits will outweigh the costs.

Part II

THE CASE STUDY

In the introduction to this book, Malcolm O'Hagan stated that "... metrication presents different planning and management challenges to each corporation, that what is right for one company may not necessarily be right for another." Recognizing that there is no single, unquestionable solution to any business problem, the case-study approach has been selected as the tool to explore the various avenues of metrication management. The case-study method encourages the reader to think in the presence of changing situations and to evaluate the possibilities, probabilities, and compromises required to reach desired objectives. In the end, the goal is reached by carefully reasoned but, ultimately, commonsense action.

Cirtem Corporation (Cirtem is metric spelled backward) is a hypothetical organization facing the major issues of metrication. Within Cirtem's organization is a situation that can be tailored to that of almost any business. For example, Cirtem is adaptable to suit the needs of large, medium, or small companies by viewing the organization as a multinational corporation, a multidivision company, a single division company, or as a group within a division. To stimulate consideration of as broad a spectrum of issues as possible, more questions are raised in the case studies than are answered in the Part III summaries.

FACT SHEET ON FICTITIOUS COMPANY:
To Be Used As The Setting For Case Method
Of Studying Metric Management

Name:　　　Cirtem Corporation

History:　　Founded in 1945. From original U.S. base, Cirtem has grown into worldwide manufacturing and marketing operations.

Location:　Headquarters in New England, U.S.A.
Overseas: Manufacturing and assembly operations in Canada, United Kingdom, Germany, Japan, and Brazil.
The Canadian and German operations have engineering departments. Marketing activities are virtually worldwide. Design and manufacturing overseas is to metric units of measure. (See Fig. 15, Cirtem Corporation Organization Chart.)

Products:　A line of walking and riding lawnmowers, and a riding tractor with attachments for gardening, snow removal, etc. The motors of the mowers and tractors have a world market.

Annual Sales: $1 Billion, of which 28% is overseas.

*Number of
Employees:* 28 000, of whom 18 000 are in the United States.

*Metrication
Status:*　Since 1970, Cirtem Corporation has been observing the world-wide metric trend ... and keeping a close watch on U.S. metrication progress. In 1974, Cirtem joined the American National Metric Council.

*New
Product:*　Now, the stage is set for Cirtem's new product program—with coordinated involvement of several overseas operations. This is

105

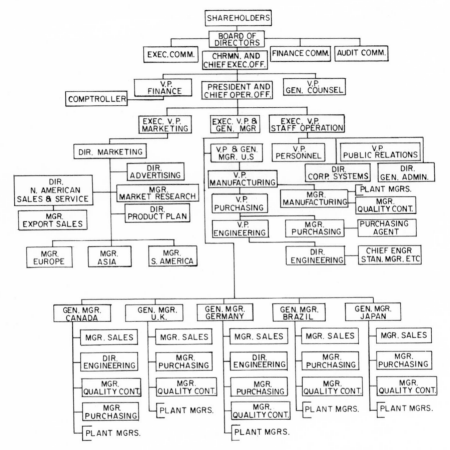

FIG. 15.   Cirtem Corporation organization chart.

to be a technically advanced product . . . new model tractor plus associated moving, gardening, and snow blowing attachments. United States operations have lead responsibility, engineering is to be shared between U.S., Canada, and Germany, and the new product line is to be manufactured initially in the U.S., as well as several overseas plants . . . with derivative products later to be produced and sold elsewhere throughout the world.

Flexibility: As a setting for discussing various aspects of metrication management, the imaginary company concept is quite flexible. For example:
Size: If the reader wishes to use a small company to develop a particular point, he can project that facet of metrication into any one operating unit of the fictious Cirtem Corporation. . . . it could be done in any one of the U.S. "profit centers" whose output is not used in the new product line.

Cast:

Charles General	President and Chief Operating Officer
James Penn	Executive Vice President, Marketing
Jim Raxzer	Vice President, Finance
John Forbes	Vice President, Engineering
Donald Reagan	Vice President, Manufacturing
Marty Becker	Comptroller
Lou Sales	Director, Marketing
Joe Bright	Director, Corporate Standards
Jack Barton	Director, Engineering Standards
Dan Holden	Manager, Manufacturing
George Lamson	Manager, Engineering Training
Fenton Riggor	Manager, Engineering
Harold Nova	Manager, Engineering
Earl Pigot	Manager, Engineering
Steve Reeter	Chief Engineer
Ken Samson	Purchasing Agent
S. I. Meeters	Metric Coordinator
and others	Purchasing, Legal, Personnel, Public Relations, etc.

COMPANY METRICATION STRATEGY
AND PLANNING GUIDELINES

Introduction

The first half of the 1970s has seen the breaking of America's metrication barrier. Company after company has faced up to the impelling need for a metric policy commitment . . . and for active movement into metric program implementation.

Throughout U.S. industry, business, labor, government, education, and the public sector one can see a gathering metric momentum . . . as the accelerating effect of interdependent metrication forces becomes increasingly apparent. The inevitable is happening: the United States is going metric.

Whether an individual company management moves by its own choice and initiative, or waits until its products and operations are caught up in metrication's rippling effect, there is need for effective management throughout the long-term transition from inch to metric.

In the 1970s, metrication is not a question of if . . . it's simply a matter of when and how. Planning the myriad of moves in making the metric transition, using logical and enlightened business strategies, and some new untried philosophies, may be helpful if not essential. Or, is the nature of the Cirtem Corporation's makeup and deduced objectives better served by minimum or no planning, leaving the "pioneering" to others and waiting for reasonable plans to unfold by turn of events, vendor and material supply, legislation, etc? If it is decided to proceed with metric implementation, when and how is it to be done?

At Cirtem Corporation, conflict arises as the various views, experiences, and objectives of the concerned management personnel surface during consideration of the early metric issues. The advocate versus nonadvocate, the fears, the enthusiasts, the problem versus opportunity opinions, and varying management philosophies all arise.

Setting the Scene

Cirtem is a medium-sized, U.S.-based company with operations in various parts of the world. Its principal product is a line of walking and riding mowing, gardening, and snow removal machines which have a novel drive design that gives them some competitive advantage. In addition, the motors of the mowers and tractors have a world market. Cirtem's overseas operations have been affected by local international movements in several countries. And now, the U.S. metric trend coupled with Cirtem's increasing world product activity—in joint engineering and manufacturing programs, as well as multinational sourcing, manufacturing, and marketing—have brought about metrication concern in the firm's headquarters in New England. It is decided that serious consideration must now be given to these concerns. Should the company develop plans for implementation? Should its primary metric impetus be based upon a "product-driven"* philosophy? Or, is a product-driven method unsatisfactory in initiating metric change because it commits action primarily in a vertical project-oriented management scheme where the horizontal management of overall company metric preparations may not be synonymous or logically coordinated?

Objective

To face the fact of gradual SI metric conversion in the U.S. and the world and decide a course of initial actions or alternatives serving the best interest of the Cirtem Corporation and its future.

---

*"product-driven": based upon desirability or economic feasibility of marketing a worldwide metric product.

The Issues

John Forbes, Vice President, Engineering, has just returned from a general manager's meeting in Germany and has called a management corporate staff meeting at New England headquarters to address the following questions:

1. Should Cirtem now proceed in its metric transition, should it proceed with accelerated actions, or should it devise new thinking?
    a. What will be the response of the key characters involved in this staff meeting: Reeter (Chief Engineer), Becker (Comptroller), Holden (Manufacturing Manager), Barton (Director, Engineering Standards), Bright (Corporate Systems), Lou Sales (Marketing Director), and others (Procurement, Quality Control, Sales, Personnel)?
    b. What will be the recommendations of the corporate staff of Cirtem Corporation?
2. If the discussion of the Cirtem Corporation staff meeting leads to suggestions for proceeding in the metric transition, what other considerations must now be addressed?
    a. Should there be a plan; if so, who has the responsibility for development; should there be a metrication committee?
    b. What is the timing for transition; should implementation start now; should there be a time limit goal?
    c. What is the real incentive: worldwide product potential; internationalize Cirtem and its standards; others?
    d. Is training necessary?
    e. Should we try to estimate costs and benefits?
    f. What further early considerations are necessary?
3. If the discussion of the Cirtem Corporation staff meeting leads to suggesting delay of its metric commitment or devising new thinking, what about the following:
    a. Impact of retention of inch/pound practices for next 5 to 10 years?
    b. What are the strategies for interfacing with the U.S. and worldwide trend toward SI use?
    c. What other comments can be made about metrication actions?

The foregoing has introduced some conditions, possible considerations, some facts, some issues, some problems and possible opportunities. Using this as a base, what are the recommendations, and solutions, (best and worst)?

Case Study 2

METRICATION COMMITTEE FUNCTIONS

Conversion to SI requires a strong commitment, sound decisionmaking, and cooperation from everyone involved in the program. Like any other investment, it will generate a profit if appropriate plans are made.

Setting the Scene

The decision has been made to go metric. Cirtem Corporation is faced with the problem of how and when. Many things go through the minds of management personnel aware of the decision. Some see it as an opportunity to clean things up and others are frightened by the prospect of a completely new measurement system.

Cirtem Corporation, like every firm, is an entity in itself. It has its own political system as well as an organizational structure. Much time could be spent arguing over committee structure and leadership. To avoid this and get directly to the heart of our discussion, let us place some constraints on our case.

Assume a corporate steering committee is formed of corporate people including the general managers from each country where there are facilities. In each operating division a metric coordinator is selected and a task force assigned to make division metric plans.

When a policy statement was issued by Charles General, President and Chief Operating Officer, saying the firm was embarking on making plans to make (SI) metric the predominant system of measurement used throughout the corporation, people began seriously thinking of metric.

Jack Barton, Director of Engineering, said, "If we tackle the problem we can save money by establishing a common language on drawings. Now our

112

people have trouble interpreting drawings from overseas plants and they have the same problem with ours." Jack Barton's voice had been heard and now he was saying, "Gentlemen, we must have adequate planning if we are to benefit the most from the transition."

Objective

To establish a mechanism to plan and manage Cirtem's conversion to metric.

The Issues

1. Who should be on the steering committee?
   a. What questions and problems face the corporate steering committee?
2. What questions must be answered by the operating level task forces?
   a. What special problems will be faced overseas?
3. What differences are there when planning for the motors sold to original equipment manufacturers and the complete machines sold to distributors?
4. What are the critical elements that could affect metrication timing at Cirtem?
5. Would a pilot program be appropriate? This would involve corporate headquarters selecting an operating division to go metric, then sharing the results with others.

Case Study 3

ENGINEERING METRICATION

Introduction

Cirtem Corporation has taken the first step in going to a world product. Engineering has been involved in planning, strategies, and guidelines. The process continued with Marketing, which has responded with marketplace evaluation. All indications point Cirtem toward launching its superior metric product and product line. (Figs. 8 and 9 illustrate Corporate and Engineering Metrication organizations.)

The Cirtem walking/riding mower comprises the retail product. The engine with its superior features is slated for the original equipment manufacturer market.

Setting the Scene

An Engineering meeting has taken place which outlined Cirtem plans to market the new product. Engineering has had, and continues to have, a heavy workload. Steve Reeter had heated discussions with engineering managers, Fenton Riggor, Harold Nova, and Earl Pigot. Each would have liked more time before jumping into a metric product. Reeter has asked for opinions and a preliminary schedule from the managers. The managers have had trouble deciding what course to take.

In a memorandum to Steve Reeter, Fenton Riggor proposed the following:

> We can produce the new product design quickly by merely adapting our present model II mower design to fit the new perpetual engine. We can be in production with this configuration until we have the proper time to redesign the sheet metal and transmission elements. Drawings can be quickly converted from

inch to SI metric units and will show dual units. You can easily see that this will yield a metric design in the shortest possible time and with the least cost.

Harold Nova has a different approach, as revealed by his memorandum to Steve Reeter:

> Although we have a very heavy schedule at this time, I feel this is the opportunity to have Pigot's engine fit into a modularized all-metric product. The schedule that I am preparing includes putting together all those suggestions that value engineering made on the model II and splitting the design segments between Riggor and myself. This is a very opportune time to try the all new solid wheel design and the new fluid drive proposed last spring. Our product will then reflect the newest innovations in mower technology. Of course, a pilot program will be necessary to prove out the new hardware.
>
> The pilot program is a must since the new transmission has not been field tested . . . .

After receiving copies of Nova's and Riggor's memoranda, Earl Pigot wrote to Reeter and asked for a meeting to settle the stated differences.

The reader should prepare an agenda for that meeting and start to solve the problems that have been posed.

Objective

To produce a basic design for a Cirtem product which can be manufactured and used anywhere in the world. Or is it?

The Issues

1. What objectives should Engineering develop in order to be in accord with Cirtem metric planning and strategy?
2. What kind of a design will be produced?
   a. A design to inch modules?
   b. A hybrid design to SI metric/inch modules?
   c. A pure SI metric design?
3. What steps, if any, must be taken by Steve Reeter (Chief Engineer)?
4. What questions does Engineering have to answer in order to produce the desired design?

Case Study 4

MANUFACTURING METRICATION

Setting the Scene

Form its original U.S. base, Cirtem has grown into a multinational company with manufacturing operations in Canada, the United Kingdom, Germany, Japan, and Brazil. The operations in Canada, the United Kingdom, and Japan were built up over the years. The plants were established in these countries to meet the growing need. The plant in Germany, however, was acquired through acquisition of a small company in a similar product line. The manufacturing plant in Brazil is the most up-to-date, as it is the latest of the overseas subsidiaries and has plans for future expansion.

The plant in Germany manufactures both the product designed at the U.S. parent company as well as products designed by their own engineering department. The management of Cirtem is aware of problems in manufacturing posed by plant location in different countries speaking different languages and working to different standards. This situation is further aggravated with servicing problems of similar products made at different locations, sold around the world. The company has lived with this problem for a number of years and has had some measure of success in resolving it by marketing and servicing in predefined geographic areas.

Objective

Establish a cost-effective metric manufacturing capability with Cirtem Corporation.

116

The Issues

Confronted with the decision taken at the meeting attended by Dan Holden, Manufacturing Manager, Donald Reagan, Vice President, Manufacturing, addressed the problem of metrication. The problems facing him were: when do we change, how do we change, and what do we change? Dan Holden's views were known to be antimetric: supply inch-based areas with inch-based products, and the overseas operations can handle metric. This is similar to what has been the practice to date.

1. Should the Brazilian plant's plans for expansion include metric machines, or inch machines, or inch machines with the capability of converting at a later date?
    a. Should this policy of machine acquisition be implemented at other Cirtem plants?
2. What should the plans be for purchasing of machine tools and accessories, measuring and gaging equipment?
    a. Should we convert?
    b. How should we convert?
3. Most people in the machine shop have personal tools. Do these tools need replacing?
    a. Who pays for their replacement?
    b. Does the company provide adequate small tools from the tool stores?
    c. Can we supply small tools and measuring equipment on long-term loan?
    d. Do we buy the tools and issue as required, or do we supply every employee with alternate metric sets to all that he has in his toolbox?
4. Will everything that we normally purchase be available in metric units?
    a. Will our purchasing people have to research and source metric items?
    b. Will the costs be any higher?
    c. What is the availability like?
5. Will our stores be capable of holding dual inventory?
    a. Will we need a new stock control system or will the existing system be adequate?
    b. Do we need to segregate metric and customary stores items?
6. Do we need to identify any of the items as metric or customary, both for components and tools?
    a. Can we extend our existing numbering system to encompass metric items to identify each with a unique number?
    b. Is there a likelihood of parts or tools getting mixed up?

   c. Can we do anything to identify them?

   d. Does this apply to customary and metric paperwork?

   e. Will it serve a useful purpose to identify this separately?

7. Is there anyplace where the use of metric units could cause safety problems (e.g., misunderstanding of units, incorrect comprehension of metric units, inadequate training, or just a relapse into the quantity appreciation of the existing inch memory banks)?

8. What will all this cost?

   a. Costs for replacing existing machines with new machines would be very high. Simple modifications to existing machines may not be adequate and may increase chances of errors and loss of accuracy. Conversion or replacement could tie the company down to one language, but chances of errors are nill. Adding conversion equipment may be costly, but could improve the performance of the machine.

   b. How old should the machine be before deciding that the cost of conversion will not be justified for the remaining life of the machine?

   c. Could the addition of conversion equipment increase the life expectancy of the old machines?

   d. Cost of machine accessories could be high, but could these costs be spread over several years?

   e. Would the existing equipment do if the company continued purchasing material in inch sizes? There may be a cost penalty in the later stages.

   f. Costs must be considered if a decision is taken to supply personal metric tools to all employees. Supply only the tools which have been identified as necessary for day-to-day operations. Have the supply available in the stores so that they can be made available as required. Have the employees buy their own tools. Maintain good industrial relations.

METRICATION EFFECT ON MARKETING

Setting the Scene

Cirtem marketing activities are virtually worldwide. The product line is essentially in the consumer market, and the distribution channel is through authorized distributors and dealers, the latter providing maintenance and repair service.

A new product has been developed. It is a technically advanced new model tractor plus associated mowing, gardening, and snow blowing attachments. The decision has been reached to make this a metric product. After a study by his marketing organization, James Penn, Executive Vice President, Marketing, has had an input in this decision, pointing out that there will be marketing advantages in foreign markets and probably no significant advantages in the U.S. In fact, there could be some objection from U.S. dealers because of need for metric parts and tools. Penn believes this domestic situation is not serious, can be overcome, and is outweighed by the advantages in design, manufacturing, worldwide interchangeability, costs, and easier access to foreign markets.

Objective

To decide whether or not to emphasize the metric aspect of the new product in the total market and/or in selected market segments.

To consider the best positive use of the metric aspect in the marketing plan.

The Issues

1. What are the key market segments?
U.S.
Canada
UK
Northern Europe
Southern Europe
Japan
Australia
South America
South Africa

2. Does the metric aspect affect optimum timing for release of the new product in each of these segments? What should the timing be?

3. What emphasis should be given to the metric design in the various market segments?

4. In order to take advantage of metrication or overcome its disadvantages, how should it be treated in each part of the marketing plan?
   a. In announcement plans and press kits?
   b. In advertising:
      • To the consumer?
      • To distributors and dealers?
      • In aids to distributors and dealers in local advertising?
   c. In publicity?
   d. In dealer sales aids?
   e. In consumer brochures?
   f. In consumer instructions—operating manual?
      • Should Cirtem provide one or more metric wrenches?
   g. In the dealer's service manual?
      • For the U.S. markets, what metric tools are needed?
      • Should Cirtem provide them free or at cost?
   h. In dealer training for service?

5. What will be the policy on spare parts backup, particularly in the U.S.?

6. Are any changes in the price structure necessary?

7. What changes must be made in the order processing system?

METRICATION OF COMPUTER SYSTEMS

Setting the Scene

Joe Bright, Director of Corporate Systems, has had several meetings during the exploratory stages prior to the decision to go metric, and several meetings since that time with his systems managers and key systems engineers and programmers. These people have been saying that they see no problems in their areas resulting from the conversion program. In fact, they have shown many of their programs that already carry both the metric and/or customary languages for linear, volume, mass, or temperature measurements.

The personnel makeup of Cirtem's systems organization is, on the average, quite young, aggressive, and intelligent. They have recently been graduated from universities where they have all had exposure to the metric system in recent years, an item which they have pointed out to Bright several times over. Their capabilities have been ably demonstrated by their solutions for very complicated problems on the job. With this background, Joe Bright has been able to accept their statements of "no problem" very comfortably in the knowledge that they understand the situation as it applies to the systems area.

There have been no specific plans drawn up for education, nor have the needs been addressed, because of previous metric exposure in school and because the systems people do not feel the need for refresher training.

The users of the systems for which Joe's area is responsible are beginning to give him an uncomfortable feeling that all may not be as good as predicted. For instance, Manufacturing Engineering completely overlooked the need to develop the kinds of information for training programs to teach engineers to "think" metric. Also, Manufacturing had thought that the cost of conversion was going extremely high in a very close time frame. However, they found that since they were already using dual dimensioned drawings, products could be made on either metric or customary calibrated equipment. In addition, conversion costs could be spread out over a longer period, during which time some costs could be eliminated either by normal replacement equipment (which would be purchased fully metric) or by having the cost be a part of a normal major overhaul program.

With some very real-life examples like these, and with increasing "noise" from users who felt there were problems the systems people were not seeing (especially from some of the users who had already had extensive metric training), Joe is beginning to feel like he is standing in quicksand.

Objective

To determine whether metrication poses problems in the computer area, identify such problems, and offer possible solutions.

The Issues

After taking action, Joe finds that there seem to be three major issues with many minor issues.

1. Manufacturing Engineering, Quality, and Manufacturing are concerned with the fact that although new part drawings are beginning to come in fully metric, the existing numerically controlled (N/C) equipment is still calibrated and equipped to handle customary dimensioned parts. However, the new N/C equipment is coming in fully metric because dual equipped machines were quoted at much higher prices.
2. N/C Engineering is concerned that the existing graphics design equipment, which is used for proving N/C tapes and for other design purposes, may not be able to handle both metric and customary program tapes.
   a. Is the issue a programming problem? A software problem? An equipment problem?
   b. Will the N/C engineers have to develop programs in both measurement languages? Will the programs accept both metric and inch cutting tools?
   c. Can the software be modified and/or supplemented in order to handle both measurement systems?
3. Production Control, Engineering, and Management in general are concerned that they may not be able to get all necessary information in

their reports, orders, bills of material, and other computer printouts in the existing formats.

   a. What are the printout problems?

      Format?

      Style (SI)?

      Symbols?

   b. What are the alternatives?

4. Finally, Joe must consider the problem as a whole and answer for himself the following questions.

   a. Who is best able to identify and define the individual problems and needs (and therefore have that responsibility)?

   b. What plans must be made and implemented?

   c. What training is needed for Systems people?

Case Study 7

METRICATION TRAINING

Setting the Scene

A metrication committee has been formed, and S.I. Meeters has been named Metric Coordinator and will chair the committee. The committee's charter states that "the committee is to serve in an advisory capacity only; it is not authorized to establish policy regarding the conversion to SI metric." This puts the committee in a rather weak position and puts the bulk of responsibility on S. I. Meeters. The Metric Coordinator is a staff member reporting to John Forbes, Vice President of Engineering.

Forbes is convinced that metric training is not a real problem, and that most needs can be met on the job. He once stated, "My engineers learned all about the metric system in college. They shouldn't require further training."

The manager of the Engineering Training Department, George Lamson, is a member of the Metrication Committee and has submitted a proposal to cover training in SI metric and multinational standards. This proposal has been shelved because of financial constraints and John Forbes' apparent lack of interest in the matter.

There are three training departments in Cirtem Corporation, not counting overseas operations. They are: Engineering Training, Manufacturing Training, and Sales and Service Training. The Engineering and Manufacturing Training departments have been in close liaison concerning metrication training, but the Sales and Service Training Department has, as yet, shown little or no interest in metrication. Their feeling is that they can "piggyback" on whatever is developed by the other two training departments.

The Standards Department, under Jack Barton, has been busy for two years writing new standards that conform to ISO and SI metric. At this time, over 50% of the necessary standards have been written and have received limited distribution. Training in the new standards has thus far been minimal and has been directed to the personnel on the new metric product only. As the product is still in development, only draftsmen and engineers have been trained in the new standards. No training has taken place regarding SI metric units because of a lack of funds and a shortage of training personnel.

George Lamson is naturally concerned about the shortage of funding and training personnel in SI metric. He sees a real problem in the near future, but has been unable to do anything about it. His proposal was shelved, yet the training will ultimately have to be done, with or without the necessary resources. He has reviewed several commercially available training programs, and has deemed all but two of them unacceptable. One of the acceptable programs is a series of programmed instruction booklets concerning SI metric units, and the other is a slide/tape program on the use of the metric micrometer and vernier caliper. He has tested the booklets on several engineers and draftsmen; the slide/tape program was forwarded to the manager of the Manufacturing Training department. Although the programmed instruction booklets were satisfactory, they were not as complete or as in-depth as Lamson felt necessary. Also, they covered a lot of peripheral information that would limit their usage when trying to train a broad cross section of job categories. Some people would be overtrained and others undertrained if these booklets were adopted. In Lamson's mind, the only solution would be to develop his own program, geared specifically to the Cirtem training problem.

Objectives

George Lamson has established the following objectives concerning metrication training:

1. To provide SI Metric and International Standards training to all affected Cirtem personnel.
2. To conduct this training at minimum cost.
3. To minimize job and product impact (lost time due to training on company time).
4. To provide training in a timely manner.
5. To avoid both overtraining and undertraining.

The Issues

1. The three relatively small training departments in Cirtem's United States operations (Engineering Training, Manufacturing Training, and Sales/ Service Training) report through different organizations; communication between them is minimal.
    a. How can communications be improved?
    b. Which training departments should be responsible for general metric training?
    c. Who handles overseas operations?
    d. How can the three groups work together to provide at least the minimum of training required for each job category?
2. Because of economic conditions (a nationwide recession), the training departments are faced with low budgets and a hiring freeze which may last for several years.
    a. Do they require supplementary budgeting for metric training?
    b. How can they get it?
    c. What about manpower to develop training programs?
    d. Should they just forget the whole thing, assuming that when the need is critical they will get the necessary funding and personnel?
3. Because George Lamson's metric training proposal was shelved, the hoped for special funding has not materialized.
    a. Should this be pursued further?
    b. How can the proposal be resurrected?
    c. Does it have to be?
    d. Is it possible to train people without additional funds or manpower?
4. Because the changeover to metric will be by product line (new products only), initial training will be vertical.
    a. Is this practical?
    b. What about new personnel on the program?
    c. What happens when a second metric product is approved?
    d. Can the training be conducted on-the-job?
    e. What are the pitfalls?
    f. Is classroom training practical?
    g. Is the time and expense of programmed instruction or video tapes justified?
    h. If so, where will the money and program writers come from?
5. Training must be timely to be effective.
    a. Can timely training be accomplished without the necessary resources?
    b. Can it be accomplished *with* the necessary resources?

    c. Can vertical training be realistically scheduled?

    d. What are the problems?

    e. Is horizontal training the answer?

    f. How can horizontal training be conducted in a timely manner?

    g. What are the alternatives?

6. Despite the training department's overtures, management is indifferent to the training problem associated with the transition to metric.

    a. How can they be won over?

    b. Do they have to be?

    c. Suppose the management agrees that metric training is critical, but cannot or will not provide the necessary resources because of other priorities?

    d. Will the impact of inadequate or no training really create a problem?

    e. Would it be more cost effective overall to "let the chips fall where they may" and let the people pick it up on the job?

    f. What about training just a few key people on the program?

Case Study 8

METRICATION COST MANAGEMENT

Introduction

Cirtem Corporation had begun to lay the groundwork for planning and eventual launching into an orderly, gradual program of metric transition. Now, however, metrication is suddenly "upon them." The whole matter has been precipitated by a top management decision committing the company to an early start on a major international metric product program.

Growth of the European market for certain Cirtem products was the main factor in this decision. The European volume potential provided a sound planning base for development of a new mower/blower product "package" that will respond to the rapidly expanding market in Europe. In addition, it will fill a gap that had become increasingly apparent (through market research) in a key segment of the North American market. As a result, Cirtem management has ordered a combination feasibility study and product planning effort to prepare for a major product program.

Because engineering design and development work is to be shared by the company's U.S. and German operations and the goal is a true "world product" concept, an early policy was adopted to the effect that SI metric measurement units are to be used. Impetus for adoption of this *metric* policy came from recognition that a single system of measurement must be used for an international product program of such magnitude. A further incentive was the requirement that a maximum number of common, interchangeable components be embodied in the product as manufactured in various parts of the world. Another factor was the emphasis placed upon pooling Cirtem's American and German technical resources—an objective that virtually dictated the standardization of a single measurement system for the joint program.

The decision to design and develop a new metric product, with Cirtem's

U.S. operations in the lead engineering role, triggered entry into a major product-oriented metric action program. Concurrently, Cirtem continued to plan and prepare for the onset of general, full-scale metrication extending over the entire product line and all facets of the firm's operations.

Thus, for Cirtem, metrication is no longer a question of *if*—it's now a matter of *how*. And imbedded in the "how" is the pervasive *cost aspect*. The scope, scale, and importance of the metric product program are of such magnitude that it is vitally in the corporate interest to conduct an efficient, well-managed, cost-effective program.

Setting the Scene

Widely differing viewpoints on these basic questions have kindled vigorous debate among Cirtem executives. Conflict has arisen over the approach that should be taken to metrication cost management. There is disagreement as to whether or not metrication cost should be forecast in detail and monitored rigorously. Different views are evident. They are being aired at the outset of Cirtem's entry into its first metric product project.

Comptroller Marty Becker advocates tight budgeting, much quantifying and reporting, close control, and strong accounting discipline covering metrication aspects of the program.

Dan Holden, Manufacturing Manager, believes production people are "conditioned to change," and that they should be allowed to handle metric cost implications by their regular accounting system for plant equipment modification and replacement—without the need to prepare additional detailed metric cost data and special reports for use outside the manufacturing area.

Joe Bright, Director, Corporate Systems, and Ken Samson, Purchasing Agent, say that for management purposes, metric should be defined as a zero budget task, and that Cirtem should carefully avoid setting in motion forces that could lead suppliers to encounter additional costs which then would be reflected in price increases for purchased parts and materials.

Chief Engineer Steve Reeter says he will leave the cost accounting methods to others. However, he wants to be assured that financial controls will not become an obstruction, since he expects the abrupt move into a major metric product activity will have an immediate, sizable impact on his engineering function.

Jack Barton, Director, Engineering Standards, has studied the metrication experience of other companies. He proposes a least-cost strategy that treats metrication as a normal business cost, using regular cost centers as control points. Extraordinary expenses are authorized through special appropriations.

In his capacity as Program Manager, John Forbes, Vice President, Engineering, and Jim Raxzer, Vice President, Finance, have held a series of meetings covering various facets of the proposed new product program. Now, they are preparing to decide how best to handle the special metrication cost issues. In this context, they are reviewing pertinent information as a prelude to issuance of guidelines for metrication cost management.

The Differing Viewpoints

Becker's View

"I don't want people to create a lot of cost information we won't use," says Marty Becker, Comptroller. He believes that for cost-effective management, a complete, detailed financial plan, accounting provisions, and control methods are essential for metrication (in view of its pervasiveness, unique character, and many "unknowns"). This would be in addition to regular product program budget and cost accounting procedures.

Becker contends that "there has to be some kind of a rational answer to the question: how much extra cost is metrication adding to this product program? It simply is unacceptable to say: 'I don't know; it's impossible to measure'."

Becker views metrication cost effects as a special "add-on" to the normal cost factors. He believes all areas should be required to forecast and track its incremental effects, and that the aggregate (extra cost) should be subject to overall cost accounting control from a central office. Here is the framework of the plan Becker would apply to metrication cost management: (1) First, a comprehensive, in-depth study would identify all items and functions affected by metrication; (2) then, metrication cost would be projected for each item; (3) the total amount would be reviewed and, after some back-and-forth "negotiations," would be approved (Fig. 16); (4) this amount becomes the separate, special funding earmarked for metrication; (5) the total figure is broken down to provide metrication budgets for all affected areas, such as Engineering, Manufacturing, Employee Training, etc.; (6) then, as the product program gets underway and expenditures begin, all areas would procure conventional items through the normal budget channels and items to be charged against the special "metrication fund" would be reviewed and authorized by the designated financial administrator for this fund (Fig. 17).

Becker is convinced that this approach would provide a degree of cost

Special Appropriation Authorization Request

August 15, 1975

GROUP/DIVISION/UNIT	PLANT	REQUEST NUMBER	AUTHORIZATION NO.
Comptroller	Finance Office	ABC-1234	98765

REQUEST	TITLE – Metric Equipment Procurement, Special		Code:
	TYPE – Project Appropriation Authorization Request		Draft

AMOUNT	INCREASE	FINANCIAL	
$460,000	FIRST 12 MOS.	AVG. RETURN ON INVESTMENT	N/A %
		PAYOUT PERIOD: YEARS	MOS.

EXPLANATION:

This Special Project Appropriation Authorization Request is being submitted to obtain authorization of funds to purchase metric equipment and to modify existing equipment from English to Metric measurement for the affected areas.

The funds requested will provide minimal equipment and tooling for engineering the new international product in the metric measurement system. The policy is to use conventional engineering and manufacturing equipment and conversion aids to the maximum feasible extent; and obtain metric instruments and machines only as necessary.

Early approval of this request is sought due to the lead time required to obtain and modify equipment. The pilot product date of x-x-xx will require equipment availability by x-x-xx and some items have x months lead time.

A formal Project Appropriation Request detailing the equipment required will be prepared and submitted for approval prior to x-x-xx.

Supporting Special Projects are being processed concurrently within affected areas. The breakdown for funds requested is as follows:

$125,000 – Engineering Dept.
275,000 – Manufacturing Dept.
60,000 – Other Areas
$460,000

PLANT		GROUP/DIVISION		CORPORATE	
REQUESTING EXECUTIVE	DATE	REQUESTING EXECUTIVE	DATE		DATE
OTHER EXECUTIVE	DATE	OTHER EXECUTIVE	DATE		DATE
	DATE		DATE	VICE PRESIDENT	DATE
	DATE		DATE	VICE PRESIDENT	DATE
	DATE	COMPTROLLER	DATE	GROUP VICE PRESIDENT	DATE
COMPTROLLER	DATE	DIV. PRES./GEN'L MGR.	DATE	VICE PRESIDENT & COMPTROLLER	DATE
PLANT MANAGER	DATE	VICE PRESIDENT	DATE		DATE

FIG. 16. Special appropriation authorization sheet.

Inter-Company Memorandum

Date

September 3, 1975

To—Name & Department   Division   Plant/Office

See below

From—Name & Department   Division   Plant/Office

M. H. Becker   Comptroller   Corporate Office

To: Messrs. _____

_____

_____

_____

_____

Subject: METRIC EQUIPMENT PROCUREMENT;
                     SPECIAL FUND USAGE

Special approval has been obtained to purchase and modify equipment
and tooling as required for metric conversion to implement the new
metric product program.

For control purposes and to insure uniformity in the acquisition of
this equipment, all commitments will require approval of the under-
signed prior to transmittal to the Procurement Office. To accomplish
this, all requests should be routed by local financial contacts to the
Budget Operations area coordinator, who will be responsible for program coordination with Budget Operations activities and obtaining approval.

M. H. Becker

cc: _____

_____

_____

_____

_____

FIG. 17. Authorization of Cirtem metrication fund outlays.

visibility that is essential for adequate control of the unknown and unprece-
dented "metric factor." He feels strongly that the result would be a cost-
disciplined, well-managed fiscal program.

Holden's Opinion

Dan Holden, Manufacturing Manager, disagrees. "How much cost control is enough? . . . How much is too much," he exclaims. Holden points out that in Manufacturing, change is a "way of life." Hence, he opposes, as unnecessary, efforts to single out metrication cost increments and subject them to special accounting treatment, entailing an extra step for review and approval of individual expenditures.

For cost-effective handling of Manufacturing's involvement with the new metric product, he contends, all that is necessary is the usual procedure for major profit center budget projection and approval, followed by realistic, hard-nosed local area implementation.

Holden realizes that metrication impact will be greater in Manufacturing than in any other area. But he has made some preliminary investigation (Fig. 18) and summarized the information obtained (Fig. 19). He believes that, contrary to the common first reaction to a proposed metric program, limited change will be needed to produce metric sizes—at least in the early stages of metrication. His findings indicate that, in general, the actual production operations in cutting, forming, molding, or casting a metric part can be accommodated without major "tear up" of plant facilities.

Some Key Questions:

   Which plants will be involved?

   What machines will be affected?

   What is the remaining useful life for each machine?

   What must be done to enable it to produce a metric part?

   Is it worth modifying, or should it be replaced?

   In purchasing a new machine, how best to provide dual capacity.

   Can newly designed metric parts be produced on existing machine tools?

   To accommodate the metric new product program, is it necessary to accelerate the normal planned replacement schedules for production machines?

   What is the current availability of metric machinery and equipment?

   What rationale will be used for cutting tools and plant consumables?

   What new measuring and inspection items will be needed for production and quality control of metric parts?

   What will be the effect on capital equipment budgets under various alternative approaches toward metric working capability?

FIG. 18. Metrication impact on manufacturing.

II. Case Study

Inter-Company Memorandum

Date

September 3, 1975

To—Name & Department   Division

From—Name & Department   Division

D. L. Holden   Manufacturing Manager   Corporate Manufacturing

Subject: MANUFACTURING METRICATION COSTS
FOR NEW METRIC PRODUCT

The following is a summary of one-time costs associated with tooling and facilitizing Cirtem plants to produce the new international product in metric dimensions.

Stamping Department $15,000

Primarily for measuring instruments, surface plate rescribes, special punches, etc.

No press or machine conversion planned at outset.

Production fixtures and machine welders to be built in inches and set to die models.

Note: Eventually, when products are primarily metric, considerable expenditure will be required for final conversion of presses, machines, and tool room equipment.

Assembly Department   85,000

Plants considered are New England, Midwest, South, and Canada.

Costs are primarily for measuring instruments, hand tools, power tool adapters, and tool room machine adaptation.

Costs by function are as follows:

Inspection and Quality Control	$24,000
Line Repair and Production	15,000
Tool Stores	8,000
Tool Rooms	38,000

FIG. 19. Metrication costs of manufacturing Cirtem's new metric product.

In terms of cost implications, capital equipment budget is one of the most important considerations, because the dollar outlay is high and costs are recovered over long periods of time. Recognizing this, Dan Holden intends to: (1) make maximum use of conventional production equipment; (2) modify existing machine tools selectively, only as proven necessary; and (3) acquire dual metric/inch capability on new purchased equipment. He knows that, in general, machine tools can be utilized to produce parts to either inch or

General Manufacturing Department $175,000

   Perishable Tools   25,000

      Primarily for drills, taps, reamers, etc.,
      estimated at 20% cost penalty.

   Spare Parts   125,000

      Includes the additional parts bank required on fasteners,
      gear sets, etc. for metric machine tools obtained from
      Europe.

      Machine tools purchased in U.S. to be bought with
      dual capability.

   Tool Room Equipment   25,000

      This provides for adapting machines for dual capability
      with dials, feed screws, or other means on a minimum
      conversion basis.

      Provision for personal tools also included.

                            TOTAL $275,000

The $25,000 cost for perishable tools is an annual recurring cost estimated at
$.40 per unit of production.

Penalty costs shown are estimates based on what is known of the product and
volumes at this time. They reflect a minimum approach of adaptation to a metric
product, rather than extensive conversion.

                                D. L. Holden

cc: _____
    _____
    _____
    _____

FIG. 19. (cont.)

metric specifications and that for an existing machine tool, the metric
product effect is mainly in terms of the "cutting" and the "measuring"
functions as related to part production.

    Holden believes he understands what is involved in developing a metric
manufacturing capability. He regards the whole question of production
machine tool modification as, ultimately, one of the most crucial in its effect
on end product quality and cost—as well as on metrication cost. He has a firm

conviction that Manufacturing people are in the best position to make the
kind of detailed decisions that are needed, and that they should retain the
freedom and authority to do so.

Reeter's Position

Chief Engineer Steve Reeter began by pointing out that Cirtem's U.S. engineering operations are now entirely inch-oriented. He noted that there will be
some added cost in designing a metric product and said that, if necessary, he
would come up with an advance estimate on the amount of extra cost
attributed to metrication.

Reeter stated: "At this point, I've had one of our people take a first 'rough
pass' at identifying metric impact in the Engineering Department. Her notes
are contained in Fig. 20. I've already been asked to approve purchase of 200
metric drafting scales. This is not a costly item, but it shows what we are
getting into."

Reeter said that detailed lists have been prepared, identifying hundreds of
engineering items that, ultimately, will be different as metrication proceeds
along its course. This survey was based on pilot groups in each major
technical department. Reeter then remarked, "We're ready to home-in and
focus our engineering metrication effort on the new project . . . as soon as we
get the 'go' signal."

"Right now," Reeter continued, "from a cost-impact standpoint, I think
the most fundamental question we face is whether we are to engineer this
metric product with our regular inch facilities (using conversion charts,
calculators, etc.) insofar as possible—or, instead, are we to view movement
of the metric product through various engineering stages as the 'trigger' for
beginning to build up a metric engineering capability? We're going to need
some management direction to issue a guideline on that."

Looking ahead, Reeter said that he anticipated one of the thorniest
corporate decisions will be whether to release engineering drawings in "metric
only" or to put computer-generated inch conversion charts on the
drawings, as some companies are doing. He noted that Manufacturing and
Purchasing should have a voice in deciding this. But, from a corporate
viewpoint, Reeter cautioned that consideration should be given to the
substantial workload and cost implications for Engineering.

Reeter concluded: "In summary, I'll leave basic cost accounting philosophy and methods to the experts. I urge that we avoid adding paperwork
and more layers of review and approval for essential expenditures. I will say
that there is just 'no way' we could operate day-to-day in a financial strait-

jacket as tight as proposed in Fig. 17. In my opinion, that degree of centralized masterminding is unworkable. Please let me know soon what metrication cost control measures are to be applied, so we can get started. I don't have to remind you this whole new product is on a rather tight schedule."

To: Steve Reeter, Chief Engineer

From: Hazel Ducket, Industrial Engineer

Subject: How to Find Examples of Metrication's Physical
         Effect Upon Cirtem Engineering

You asked me to give you examples of the tangible effects of metric conversion on Engineering. What physical things would be different in a metric engineering operation?

To answer this question, one should look for metrication's physical effects in the product engineered . . . the facilities . . . and the equipment. Every department should be studied—because, to some degree, every department would be affected.

The review should cover all laboratories . . . drafting rooms . . . shops . . . garages . . . offices . . . Technical computer center . . . and the test facilities.

And, after having done all that for "soft conversion" (change in measurement system) . . . we should then repeat the entire review in terms of "hard conversion" (this is "soft conversion," plus introduction of metric standards that change the sizes of things).

Also, to sense the full physical impact upon Engineering, think in terms of a wholesale change in measurement units and the devices used to measure, record and compute various quantities throughout the entire design, testing, development and releasing activities.

Keep in mind that different (metric) terms are to be used for linear dimensions, area, volume, mass, temperature, and force. The new units will be: meter, cubic meter, kilogram, degree Celsius, and newton. Area is square meter . . . density, kilogram per cubic meter . . . pressure and stress, megapascal . . . torque, newton meter . . . power, kilowatt . . . viscosity, pascal.

Length will be expressed in millimeters . . . engine power, kilowatts . . . fuel and oil quantity, liters . . . weight, kilograms.

The everyday use of measurement units in Engineering is deep rooted and pervasive A sweeping change to a different system of measurement units would have far-reaching consequences, affecting literally thousands of physical things in Engineering. On the next page, you'll find some examples of the kinds of things that would be different in a metric tractor/mower/blower engineering activity.

As instructed, we've begun a pilot study. This will be a step toward identifying the metric impact. Already it is apparent that the list of items affected by metrication will be sizable and costly. Early decisions are especially important, because they will set a pattern for later purchases.

FIG. 20.   Metrication's effect upon Cirtem engineering.

Examples of Things Affected by Metric Changeover in Engineering

Layout & Precision Inspecting Equipment: Surface gages . . . vernier calipers . . . micrometers . . . gage blocks . . . thread gages . . . snap ring (clamp load) indicators . . . surface plates.

Laboratory Equipment: Pressure gages . . . radius gages . . . bore gages . . . torque wrenches . . . tensile test machines . . . hardness testers . . . thermometers . . . temperature gages . . . weighing scales . . . linear measuring scales . . . flowmeters . . . tools.

Drafting Rooms: Drafting machines (with attached scales) . . . metal drafts (grid lines 100 mm) . . . reference documents (relative to hole sizes, sheet metal gages, tolerances, etc.) . . . templates . . . sweeps (French curves) . . . and possibly different paper sizes.

Shops: Have not yet determined what physical things would be different in a metric machine shop . . . metal shop . . . die model shop . . . wood shop . . . and paint shop.

Offices: Full tangible effect not yet assessed—but, depending upon the nature of work performed in a particular office, there would be varying physical effects from metrication. Rulers obviously would change—and, if metric paper sizes are adopted, filing cabinets could be affected.

Technical Computer Center: Thinking first in terms of "soft conversion" and then "hard metric conversion"—metrication's potential impact has not been determined . . . but physical changes probably would occur to some extent in numerous items ranging from digitizers to tape and disc dimensions. (Software programs will be affected, too.)

Test Facility: Here, examples of physical change come to mind readily be realizing that we are hypothesizing a complete change in units for the quantities measured and the things used to measure them.

Threaded Fasteners: Threaded fasteners should be singled-out for special attention in assessing metric impact. When metric fasteners are adopted, we'll experience a great proliferation of releases and part variety—plus innumerable tangible changes in such things as drills, taps, washers, wrenches, sockets, etc.

Closing Observations:

1. In the foregoing attempt to give examples of physical change metrication would cause in Engineering, we're seeing only the "tip of the iceberg."
2. We should keep clearly in mind that this memo deals only with physical, tangible things. Yet, in a large Engineering operation, the intangible impact would be equally great. Apart from the training cost (more than $100 per salaried employee), the effects would include a significant loss in efficiency and productivity during the prolonged transition to a new "measurement language" . . . and a substantial increase in errors (7% error in dual dimensions), mistakes in calculations, and computer programming inefficiency.
3. Looking at what has been written in this memo, I find it is not very impressive! There is no single "big item"—no dramatic, attention-getting "problem." Apparently that's because a thousand "small things" somehow are not as impressive as one "big thing" . . even though, in aggregate, the thousand small items may far outweigh the effect of one single big item.

FIG. 20. (cont.)

Barton's Proposal

Jack Barton, Director, Engineering Standards, began by saying, "to speed up the process of getting a consensus on the basic approach we should use for metrication cost management, Joe Bright, Director, Corporate Systems, and I talked this over. We combined our ideas and would like to present a proposal that we think satisfies the common goals we all have in mind for devising a least-cost metrication strategy applicable to Cirtem's first U.S. metric product program."

Barton reported these highlights pertinent to cost control:

1. Preliminary assessment of metric pertinent to cost control:
   use sampling technique, estimate unit cost effect for cross section of items; then extrapolate for total.
2. Using this information as input, reasonable and uniform interpretations of metrication's financial requirements will be covered by individual budgets for all areas.
3. No separate budget is provided for metrication costs but provision (procedure) is made for special budget supplements when justified for extraordinary outlays.
4. As the product program progresses, each profit or cost center is a control point, responsible in the usual manner for satisfactory performance to budget.
5. No special fiscal practices need be invoked; accounting will operate under regular controls and procedures.
6. Local managers will be expected to fulfill all of their responsibilities, including metrication, within budget.
7. The overall "control" is that the profit plan must be successfully achieved.

Barton said he and Joe Bright recommend this as the framework of a sound approach. He then commented that an overall company metric plan does, of course, include suitable provisions for metrication cost control. This portion of an Australian company planning document is illustrated by Fig. 21. Barton got it from a metric committee contact at the trade association. It shows a fragment of information one company used to make the compromise between tight control, loose control, and no control.

Barton commented that Purchasing Agent Ken Samson also concurs in the foregoing proposal. He then said: "If you will agree, we could use a similar approach at Cirtem to:

1. Generate some "ballpark" metrication cost figures to at least satisfy ourselves that we had a general idea regarding the cost, and enable us to respond to upper management's interest in an overall estimate.

2. Provide sufficient data for preparing metrication cost control guidelines and charting consistent application throughout the company.
3. Assure the recording of data for uniform adherence to the general guidelines.
4. Develop some confidence that we are handling metric cost impact in a business-like, efficient manner."

Barton stressed the importance of having divisions and profit centers address metrication cost control in their operating plans. He concluded by pointing out that an inherent feature of this approach is that the first metric product will not bear a disproportionate cost penalty from its role as the "leading edge" of Cirtem's movement into metric transition.

Three different views of metrication cost management are expressed in the four foregoing statements. Someone must choose between them. For Cirtem, the decision will be made by John Forbes and Jim Raxzer, vice presidents for Engineering and Finance.

They know Cirtem is only now mounting the lower slope of Cirtem's rising "metrication learning curve." And they recognize it is important to "start up the right path" toward effective administration of metrication costs. Certainly everyone would agree that clear justification should be required before imposing additional financial duties on people in Engineering and Manufacturing.

The basic issue is clear. Forbes and Raxzer must adopt a metrication cost management policy that will provide appropriate financial control and yet not deter the product program. The problem is real: a metric product design is going to start soon in what, up to the present time, has been an entirely inch-oriented Engineering, Manufacturing, and Supply system.

How can Cirtem best handle the extra costs that will be peculiar to this program because it is metric? Should management go along with Becker's idea that tight controls are needed because of the program's unique character? Or would it be better to endorse the opposite view, as advocated by Holden? Perhaps there is merit in the Barton/Bright proposal that metrication cost be left to control by regular accounting methods, with decisions made at the operating level and some flexibility to handle exceptional metric expenses.

Excerpt From ...

... ABCDEF Corporation Plan for Metric Conversion

Purpose: To establish the policies, procedures, and practices to be used in converting ABCDEF Corporation to predominant use of the SI metric measurement system.

Section 7 METHOD

7.20 Cost Accounting Aspects

7.21 Cost Accounting is to convert fabrication, assembly, and commodity control cards for all products in phase with the Product Engineering and Manufacturing Engineering conversion programs.

7.22 In addition, Cost Accounting is to:
  a. Revise purchase order prices where changes of dimension lead to price change.
  b. Provide Comptometer service to Finance Division on all forms of change.
  c. Introduce a system to isolate product cost variants due to metric conversion activities, from other "economic" product cost variations.
  d. Introduce a system of recording cost of, or time spent on, metric conversion activities in all departments to support tax negotiations.

Section 8 FINANCIAL & BUDGETARY IMPLICATIONS

8.11 The government announced "adoption" of the metric system of weights and measures. There will be no legal enforcement of the change, but a variety of government measures will induce and later force the changes upon industry. The term "adoption" was used to divert costs involved from becoming a charge upon the Government. The government position is that costs of conversion must be borne by those incurring them.

8.12 Nevertheless, since the costs are expected to be significant, the Company will request the Chamber of Manufacturers to approach the government for some form of tax relief. A request for a direct grant would not be successful, because it would conflict with government policy. Therefore, the alternatives are:
  a. A tax concession similar to the export market development rebate provided under Section 160AC of the Income Tax Assessment Act. In that case, tax concessions would provide rebate for a portion of the expenditure.
  b. Failing the above, request that a special deduction be allowed for capital expenditures for income tax purposes.

FIG. 21. ABCDEF Corporation plan for metric conversion.

8.2   Operational Factors

8.22   The predicted workloads and estimated costs of company metric conversion
       activities are summarized in Table No. 1. The following comments refer to that
       Table:
       8.22.1   Identifiable costs are estimated at $2 945 000. This figure does not in-
                clude the inevitable but less readily identifiable temporary inefficiencies,
                which could cost and additional $75 000. The total cost, therefore, is on
                the order of $3 020 000.
       8.22.2   Because much of the work will be absorbed into normal staff duties,
                actual expenditures will be some $220 000 less than that total cost. Thus,
                direct cash outlay on metric conversion is estimated as $2 800 000.
       8.22.3   The $2 800 000, which is mainly for conversion of plant and equipment,
                assumed physical conversion would be required, as for example, replace-
                ment of lead screws and gears. The planned investigation of individual
                machines to ascertain the extent of conversion required to gain the
                simplest, most economic method of metric operation, is expected to
                reduce the estimate substantially.

8.23   The company metric conversion workload, estimated as 155 000 manhours, will
       be spread over a number of years. Metric conversion activities are, therefore, to
       be included within annual departmental manning levels and activities. Additional
       staff will not be authorized specifically for metrication, unless applications are
       justified and approved in accordance with standard procedures.

8.24   Using the system established for that purpose, all departments are to record total
       time spent on conversion activities, to provide detailed supporting data for
       possible tax considerations.

Table 1

The total cost of ABCDEF Corp. metric conversion activities is estimated in the following tabulation. Estimates are based upon the company's standard labour rate. Items
listed represent work that would not be necessary, but for the need to convert activities
to the metric system. The work, and so the expense will be spread over a number of
years—and much of the work will have to be absorbed into normal duties . . . therefore
direct cash expenditures will be less than the total indicated cost of conversion activities.

PRODUCT ENGINEERING

	$ (Aust.)
Convert Material Standards and Engineering Specifications	20 250
Prepare Metric Standards Parts Book	26 250
Revise Drafting Standards Manual	3 750
Releasing Effort to Convert Bulk Materials	2 250
Revise Carryover Drawings Affected by Changed Gauges and Standards	60 000

FIG. 21(cont.).

	$ (Aust.)
Convert Test Procedures and Test Affected Parts	3 750
Convert Engineering Book and General Vehicle Specifications	3 750
Design Engineering Support to Materials Engineering, etc.	30 000

MANUFACTURING ENGINEERING

Convert Blank Preparation, Process Planning, Mtl. & Press Load Documents	26 250
Convert Tool & Die, Press, Fixture, Welding, and Stock Tooling Standards	32 625
Revise Plant and Facility Standards	750
Manufacturing Engrg. Support for Plant Conversion Activity	7 500
—First annual project (metric measuring equipment only)	200 000
—Convert plant, equipment and workers tools	2 250 000

SERVICE DEPARTMENT

| Revise Service Manuals, Owners Manuals, Pocket Guides, etc. | 5 625 |

PARTS & ACCESSORY DIVISION

| Specification and Call-Up of Changed Replacement Parts | 1 875 |

PURCHASING

Requote on Carryover Parts	
Purchasing Research Investigation of Price Increases	
Revise Quotation Charts	112 500
Reissue Purchase Orders, Drawings and Specifications	
Requote Non-Production Supplies	
Reissue Non-Production Purchase Orders	

PERSONNEL

Prepare Training Material	750
Present Training Courses	750
Consultative Training Assistance to Departments	750

MARKETING

| Sales Training | 750 |

PRODUCTION CONTROL

| Records Conversion and Dual Running; | |
| Steel, Trim, Non-Productive Items | 4 800 |

FIG. 21. (cont.)

$ (Aust.)

FINANCE

Revise Data Processing Programs	5 625
Revise Internal Forms	1 500
Revise Fabrication, Assembly, and Commodity Coding Cards	6 750
Revise Purchase Order Prices	6 750
Provide Comptometer Services to Finance Division	1 500
Print Converted Standards and Manuals	5 625
Print Bulletins, Training Materials, etc.	1 875

MISCELLANEOUS COSTS

Time Spent by Staff at Conversion Briefings	3 750
Time Spent by Staff at Conversion Training Courses	11 250
Time Spent by Metric Committee Members on Metrication Activities	28 500
Travel Expenses, to Attend Industry Metric Committee Meetings	3 750
Paper and Binders for Converted Documents	21 250
Purchase Metric Design Handbooks, etc.	2 500

COMPONENT MANUFACTURING SUBSIDIARY CO.

Total cost impact of metrication on all operations	50 225

Grand Total	$2 945 000

Note:
 Factors not costed above include:
  Loss of design and production efficiency
  Additional work due to
   Soft conversion of SAE, ASTM, and other documents
   The need, initially, to consider the implications, and, later to adjust individual activities to metric
  The necessity to maintain both metric and imperial Standard Parts Books and other design data for up to 10 years.
  Product cost variations attributable to metrication.

 In aggregate, these factors could represent a loss of efficiency of 2% . . . this has not been qualified.

FIG. 21 (cont.).

Objectives

Through application of regular accounting procedures, it is desired to generate sufficient financial information to:

1. Enable effective cost management for Cirtem's first international metric product program.
2. Isolate discrete metrication costs as necessary to: (a) provide financial control over this aspect of the program; and (b) create records that may be useful for internal management purposes, or advantageous for various tax considerations.
3. Overall, to achieve the least costly path to a metric product.

The Issues

1. Are metrication costs negligible or prohibitive?
   a. Can they be accurately measured?
2. Should metrication costs be isolated and tracked?
   a. Or is cost measurement counterproductive?
3. Should metrication costs be budgeted as a separate line item?
4. For accounting purposes, should the cost of metrication be treated the same as other cost aspects of product programs?
5. How closely should metrication costs be forecast, measured, monitored, and controlled?

Part III

CASE STUDY ISSUES IDENTIFIED

It is emphasized that the summary of issues presented here applies to the specific case of the fictitious Cirtem Corporation, as discussed by the identified moderator and this particular group of participants. All of the people did not agree on all of the points; however, the items listed do represent a group consensus. With a different moderator and different participants, other issues may have surfaced. What is presented here is typical of the major points that would normally arise. The statements made in this section are intended to be thought-provoking ideas rather than definitive answers.

Saul Alford
Metric Coordinator
Argonne National Laboratory
Argonne, Illinois

William E. Brueckner
Program Planning Specialist
Xerox Corporation
Webster, New York

Robert A. Bruening
Manager, Engineering Services
Chicago Bridge and Iron Company
Oak Bridge, Illinois

Douglas R. Burke
Head, Standards and Materials
Eng. Dept.
Bell Telephone Laboratories, Inc.
Holmdel, New Jersey

Phillip A. Dressing
Metric Planning Group
Goulds Pumps, Inc.
Seneca Falls, New York

Frederick F. Forte
Section Head, Corporate
Engineering Dept.
The Coca-Cola Company
Atlanta, Georgia

C. S. Hines
Mgr., Chemical & Plastics Standards and
Corporate Metrication Coordinator
Union Carbide Corporation
South Charleston, West Viriginia

Wayne A. Hoffman
Standards Engineer
Giddings & Lewis Machine Tool Co.
Fond du Lac, Wisconsin

Donald M. Kladstrup
Director, Long Range Planning
Distribution Division
Eastman Kodak Company
Rochester, New York

Jim MacKenzie
Special Projects
Ralston Purina of Canada, Ltd.
Rexdale, Ontario, Canada

Robert G. Michel
Manager, Engineering Systems
Allis-Chalmers Corporation
Milwaukee, Wisconsin

Irving R. Miller
Manager, Quality Control
Allen Tool Corp.
Syracuse, New York

Reynold F. Missall
Group Package Engineer
Avon Products, Inc.
New York, New York

William G. Moore
Vice-President, Engineering
Lilliston Corporation
Albany, Georgia

Kenneth L. Morrison
Metric Coordinator—Brass Group
Olin Corporation
East Alton, Illinois

Richard B. Neal
Director, Resources and Patents
Rexnord, Inc.
Milwaukee, Wisconsin

Roy L. Pollack
Chief Engineer
LCN Closers
Princeton, Illinois

Philip P. Scalia
Manager, Drafting Services
Taylor Process Control
Division of Sybron Corp.
Rochester, New York

George W. Schiff
Engineering Planning Director
Raytheon Company
Lexington, Massachusetts

Gene D. Sickert
Senior Manufacturing Analyst
Machinery Group Headquarters
FMC Corporation
Chicago, Illinois

M. E. "Jerry" Smith
Project Engineer
Central Engineering Laboratories
FMC Corporation
Santa Clara, California

Ronald F. Strahan
Coordinator—Metric Conversion
Outboard Marine Corporation
Waukegan, Illinois

James G. Van Oot
Metric Information Coordinator
E. I. du Pont de Nemours & Co., Inc.
Wilmington, Delaware

COMPANY METRICATION STRATEGY AND PLANNING GUIDELINES

Summary

Congressional action and a sizable (28% of $1 billion) overseas market leave Cirtem with no alternative but to start planning for metric conversion. Additionally, timing metric planning with the introduction of a new product line presents an ideal opportunity for Cirtem to get in step with the rest of the world. Cirtem's management should plan with an eye toward the future, realizing that the environment in which metrication will be implemented will be different than the environment in which it was planned. Cirtem must do today's job while preparing for the future.

Who will lead Cirtem toward adopting metric units? It will probably be middle management's task to convince top management of the advantages of the changeover. However, there will probably be a difference of opinion on metrication among middle management. For example, in many businesses and industries, technical people (engineers, R & D personnel, etc.) are in the forefront, with the marketing people being dragged in kicking and screaming. However, there is a market drive for metrication in some industries (steel, for example), and the marketing people will lead the way in these instances.

Different levels of agreement on metrication will be evident at John Forbes' (Vice President, Engineering) staff meeting. While there will be a general consensus for metrication, the participants will express the concerns of their respective departments:

Steve Reeter (Chief Engineer) will be concerned about metrication's effect on design resources (standards, materials, etc.); he will also want to know if Cirtem plans to adopt the SI metric system, and how conversion of the U.S. plants will be coordinated with Cirtem's foreign divisions.

151

Marty Becker (Comptroller) will want to know how much metrication is
   going to cost, and when its benefits will start outweighing its costs.
Dan Holden (Manufacturing Manager) will want to know which workers
   will have to be retrained, and what machines and tools will have to be
   converted or replaced.
Lou Sales (Marketing Director) will want marketing to play a major role
   , in setting the time frame for conversion.

However, the important thing to remember is that metric conversion is
neither a technical nor marketing decision; nor is it a matter of metrication
for metrication's sake. It is strictly a *business* decision. Once top management
realizes that metriction is a business opportunity, that short-term costs will
be offset by long-term opportunities, they will move quickly to plan for and
implement metric units.

Convincing Cirtem's top management will take some doing. They will
probably not understand what all the fuss over metrication is about. ("Is
metric something you do to people or with them?" is a typical response.)
To them, metrication is simply a matter of changing the labels on Cirtem
products.

Above all, they will want to know how much metrication is going to cost.
If they are unfamiliar with metric conversion, Cirtem's top management
(especially Marty Becker, the comptroller) will want exact cost estimates.
They will not realize that you can spend more time and money tracking costs
than you can on metrication itself. And hopefully, they will realize that what
appear to be costs now can actually become benefits in the future.

Top management will ask many other questions. Who is going to be
responsible for the training which will be necessary to make the changeover.
How do our competitors feel about metrication and what are they doing in
this area? Are our markets, both U.S. and foreign, ready for a metric
product? What is the government doing and planning to do about metric
conversion? Should the new product line be metric or should we wait for the
next product? Or the one after that? Should it be a metric product commitment or a total metric commitment?

Once they get over the "why can't the rest of the world use the inch?"
syndrome, Cirtem engineers and shop people will begin asking questions.
Should the SI metric system be adopted? Should we soft- or hard-convert?
Can our suppliers provide the necessary parts and materials? Does the shop
have the capability to produce metric products? How will international
standards affect the product and product planning? What is conversion going
to do to our computer programs?

These are all legitimate concerns, and it will be necessary to deal with them as Cirtem adopts metric units. However, solving these problems will follow, not precede, Cirtem's commitment to go metric. This commitment has been made, and Cirtem is ready to form a corporate committee to investigate metrication and develop a corporate metric strategy.

Decision-Maker's Metric Management Checklist

1. Metrication: Preplanning Considerations
   a. Recognize the need for in-depth planning at all levels
   b. Weigh metrication costs versus benefits, the problems it presents versus the opportunities it offers
   c. Determine whether the market (U.S. and foreign) will support metric products
   d. Determine the availability of supplies and metrication's effect on suppliers
   e. Determine engineering and manufacturing metric capability and resources
   f. Determine the extent of employee training metrication will necessitate
   g. Determine metrication's impact on overall company operations
   h. Examine industry, national, and international standards
   i. Assess the metric policies and activities of competitors
   j. Be aware of governmental (local, state, and national) metric policies affecting your company and products
   k. Obtain top management support and coordination for the company policy toward metrication
   l. Implement on the basis of what is right for your company
2. Metrication: Planning Requirements
   a. Policy statement
   b. Overall plan
   c. Metric coordinator(s) and committee(s)
   d. Optimum conversion timing
   e. Cost estimates
3. Metrication: An Action Plan
   a. Obtain a metric commitment
   b. Policy statement (including corporate strategy)
   c. A plan (including selection of who is to be responsible)
   d. Considerations: costs, training, marketing, engineering and production, standards, and competitors

Case Study Results 2

METRICATION COMMITTEE FUNCTIONS

Summary

Cirtem is going metric. When and how they choose to do so will depend on the corporation's particular circumstances. Although looking at the experience of others is helpful there are no clear-cut answers. A corporate metric committee will have to decide what is right (and what is wrong) for Cirtem, and plan accordingly.

The chairman of the corporate committee should wear the corporate hat; that is, he should have a broad view of metrication and what it will mean for Cirtem. He should be committed to metrication and should not view it as a peripheral function. Cirtem's metric committee chairman should be a senior company officer, someone who commands respect and has the authority to get things done. He must also be an effective liaison and be able to work at all levels of the company. Since Cirtem is a consumer-oriented company, top management might want to consider James Penn (Executive Vice President, Marketing) for metric committee chairman (if he satisfies the other criteria, that is).

How many members should Cirtem's corporate metric committee have? Who should be represented? The committee should be large enough to have representation from all important divisions of Cirtem, but small enough to get the job done. If the corporate committee is too large, it will prove unwieldy, and will not be able to function effectively. Since Cirtem is a centralized organization (it produces one basic product), a 6-8 member committee would be best.

Membership on the corporate metric committee should include representatives from finance, marketing, and foreign operations. Foreign representation is necessary because Cirtem's design and manufacturing overseas is to metric units of measure. Including foreign engineering or marketing representatives

154

will shortcut lines of communication, which is important if Cirtem is to learn from its overseas experience. And there is much to be learned from overseas production because it has been, in effect, Cirtem's pilot metric program.

The corporate committee will develop general guidelines and policy direction for converting in a cost-effective manner. Their final report will be submitted to Cirtem's top management for approval. The next step will be to form metric committees and appoint metric coordinators at the plant level, where the plan will be implemented.

Cirtem's corporate committee should be involved with these plant level committees. This does not necessarily mean knowing what is happening on a day-to-day basis, but rather providing guidance to solve the many small problems that metrication will cause. Cirtem's decision to go metric will affect all employees and, by being present at the plant level, the corporate committee will be able to handle the problems and minor details of Cirtem's metric program promptly and effectively.

One final note: just as on the corporate level, there are no absolute maxims for metric conversion at the plant level. Operations in foreign plants will differ from U.S. plant activities; each plant must depend on different suppliers and meet the needs of different markets. The structure and function of Cirtem's plant level metric committees should be based on these circumstances.

Decision-Maker's Metric Management Checklist

1. Metric Committee: Organization and Structure
   a. The committee:
      - must have a broad understanding of metrication and its probable impact on the company
      - must include representation from all key company functions
      - must be small enough to study the problem effectively and implement the plan
      - must have top management support
   b. The Metric Coordinator:
      - must have a broad understanding of metrication and what it will mean for the company
      - must command respect and have the authority to implement the plan
      - must be able to act as an effective liaison
      - must be able to work at all levels of the company

2. Metric Committee: Functions
    a. Develop metric policy and guidelines
    b. Establish lines of communication at all levels of the company
    c. Keep employees aware of metric planning through a low-key approach
    d. Determine optimum timing for the transition
3. Metric Committee: The Timing Function
    Timing will depend on:
    - inventory on hand
    - lead time on tooling and gages
    - availability of purchased components, raw materials, and other supplies
    - market conditions
    - original equipment manufacturer transition plans
    - availability of international standards

ENGINEERING METRICATION

Summary

Cirtem is starting a new product line, and has decided that the tractor and attachments will reflect some degree of metric implementation.

The new tractor will have to meet the needs of Cirtem customers in the U.S., Japan, Germany—wherever they are. More often than not, the customer could care less whether the product is inch or metric. He just wants the tractor to do what it is supposed to do; he also wants good servicing capability, whether he does it himself or goes to a Cirtem dealer. Cirtem's engineers should keep these things in mind as they design the new product.

One of their first considerations will be to choose from the three methods of conversion: pure metric, hybrid inch/metric, or inch module/soft-converted. Product lifetime is an important factor in this choice. Because the new product will go into the metric era, it will be to Cirtem's advantage to convert as much of the tractor and parts as possible during initial design and production. Also, design modularization would ease Cirtem into metrication gradually and should be considered. Cirtem might also decide to produce two products, an inch one for the U.S. market and a metric product for the foreign market.

There are many questions to be answered before Cirtem's product design is finalized. What is the difference in cost among the design alternatives? Can Cirtem get the material to produce the product it designs? Can its shops manufacture the product its engineers design? If a metric product is developed for the U.S. market, how will Cirtem handle servicing? How will engineering codes and standards, both in and out of house and in and out of the country, affect the product? How will legislation and government regulations affect design? These problems lead to an important planning consideration: Cirtem should prepare for a heavier design load.

Finally, if Cirtem decides to soft-convert, only engineers should put the conversions on the design and product. This will help to maintain design integrity and prevent possible problems in rounding the conversions.

Decision-Maker's Metric Management Checklist

1. Engineering Metrication: Strategy
   a. Engineering strategy should be based on overall company objectives.
   b. Determine the metric units best suited to the company's needs and adopt them as the official measurement system.
   c. Time the implementation in Engineering with the changeover in other company operations, especially Manufacturing and Training.
2. Engineering Metrication: The Design
   a. Select the design (pure inch, hybrid inch/metric, or pure metric) which will be best suited to the product and the market.
   - The alternative of an inch product for the inch market and a metric product for a metric market should be considered.
   - Modularization will ease the transition from hybrid to pure metric over the years.
   - Depending on the market, it can be advantageous to identify metric portions of the product.
   b. Determine the engineering codes and standards that will be affected by metrication and the role other codes and standards will play during the transition.
   - Revise company design manual.
   c. In order to maintain design integrity and prevent confusion in rounding, only engineers should put conversion figures on the design and product.
   - Establish a companywide measuring policy.
   - Justify the use of dual dimensions in writing.

MANUFACTURING METRICATION

Summary

Mention metric conversion and the first reaction you will get is "how much is it going to cost?." This is no exception with manufacturing metrication, where there will be two sources of costs: conversion or replacement of machine tools; and employee training, which will be discussed later.

When and where should Cirtem manufacture their new product line and what quantity should be produced? Although the answers to these questions depend on market conditions, the obvious answer would be to manufacture the tractor and parts overseas where production is already in metric units.

However, the obvious answer is not always the right one. Investigation may reveal that Cirtem does not have the technology or the capability to produce the new product line overseas. The important thing to remember is that Cirtem has made the commitment to go metric. If Cirtem decides to manufacture the metric product overseas at this time, sooner or later the company will have to manufacture the product in the U.S. Their decision should be based on economics, the market influence, and other variables.

The importance of the new product line and its anticipated lifetime are two more considerations. Depending on whether a high or low value is placed on the tractor, Cirtem may decide to either convert its existing machine tools or replace them with new, high quality ones. The same is true with the expected product life. Cirtem will have to decide on a cost-effective approach: to use converted machine tools until they become obsolete or to purchase metric machine tools that will manufacture the product on into the metric era.

Manufacturing a metric product will cause problems with supplies, inventories, and space. Cirtem cannot outpace the metric capability of its suppliers. On the other hand, it must not wait for its suppliers to take the initiative.

159

There must be a dialogue, with Cirtem bearing the burden of demonstrating to its suppliers the benefits they will realize as a result of Cirtem's decision to go metric.

Balancing inventories with the timing of metric production is another problem that needs to be attended to. Cirtem must decide not only how long it wants to continue marketing an inch product, but also how long it wants to have a servicing capability for that product. If servicing is to be for any length of time, storing inch parts will create problems.

If Cirtem decides on dual production (manufacturing an inch product and a metric one), its problems will be doubled. Dual production will require two sets of drawings, supplies, tools, etc., all of which will occupy twice as much space. Machine shops might have to be rearranged and, depending on the circumstances, Cirtem may want to segregate metric production and stores.

Replacing personal tools is another problem, serious enough to have delayed congressional approval of metric legislation. Labor unions are concerned that their members will have to foot the bill for replacing inch tools with metric ones. Statements about "letting the costs lie where they fall" have not helped matters much.

The problem can be solved in a number of ways. A tool replacement policy can be based on whatever management can negotiate with the union or on "whatever will keep the troops down in the shop happy." Or Cirtem can continue with its current practice, with either management, the employees, or a combination of the two supplying the tools.

Cirtem will more than likely follow the example set by other U.S. companies: supplying small tools to the worker on long-term loan and making larger tools available in the tool crib. In many cases, the long-term loan will turn out to be a giveaway, but these costs can be minimized by supplying only the tools that are necessary to do the job. Also, many workers who have built up a collection of tools over the years may want to buy their own metric tools. Recalling the loaned tools periodically will accomplish two things: it will keep track of the tools; and ensure that they are kept in proper working order.

A change in the work rate (pieces per hour) or the pay standard (wages per peice) will be two other union concerns. Again, Cirtem will have to go with whatever can be negotiated. Finally, a successful metric awareness program should eliminate some of the ill feelings the union might have about metrication.

However, no matter how successful Cirtem's planning is, there will inevitably be unforeseen production problems. For this reason, the first run of the new product should be carefully monitored so that these problems are identified and the necessary corrections made. Increased production time, confusion caused by dual dimensioning, different maintenance requirements, and new methods for inspecting incoming materials are just a few of the areas where problems can (and will) occur.

Decision-Maker's Metric Management Checklist

1. Manufacturing Metrication: Tooling and Equipment
   a. The decision to convert or replace inch machines and tools should be based on:
      - machine life expectancy (metrication costs are not incurred if a machine is scheduled for replacement shortly)
      - delivery time for new machines
      - delivery time for conversion equipment
      - time required to convert existing equipment
      - life expectancy of the equipment following conversion
   b. When acquiring new machines and tooling, purchase equipment with metric capability.
   c. New machines should have dual readout capability.
      - Machines should have metric readout only as metrication progresses
   d. Supply personal metric tools to workers on long-term loan and provide larger metric tools in the tool crib.
      - Supply only those tools necessary to do the job.
2. Manufacturing Metrication: Facilities
   a. Stock only those inch items required to complete existing contracts or authorized production runs.
      - Stock inch parts for only as long as you want servicing capability.
      - Promptly dispose of obsolescent or surplus slow-moving inch items to make room for metric components and materials.
      - Initially, some dual stocking will be necessary.
      - Segregate metric stocks from inch stocks.
   b. Clearly and uniformly identify metric components, parts, etc., especially if the metric item is not obviously different from its inch counterpart.
      - Use a color or number code.
   c. If necessary, rewrite planning sheets, cutting lists, and routine instructions.

  - Write inspection and test specifications for new products in metric terms.
  d. Obtain metric standards for manufacturing departments with standards rooms.
    - Initially, it will be necessary to maintain customary and metric standards.
  e. Look for potential safety hazards caused by confusion or misunderstanding of metric units and terms.
3. Manufacturing Metrication: Timing
  a. Coordinate company metrication with the conversion timetables of suppliers and subcontractors.
    - Receive firm commitment dates from suppliers and subcontractors for the availability of metric machines and tooling, components, raw materials, etc.
  b. Give the company buyer firm planning dates for the introduction of metric products so that he can:
    - locate stocks of metric items and raw materials
    - locate and evaluate new sources of supply
    - arrange for new tools to be ordered, secured, and tested
    - secure batch quantities of required metric items and raw materials
  c. Allow lead time for employee training.
  d. Plan for increased delivery times metrication will cause.
  e. Monitor initial metric product run to identify and correct problems.

METRICATION EFFECT ON MARKETING

Summary

Cirtem has made a firm commitment to go metric. It must make the best possible use of this commitment in its marketing plan. Cirtem's consumer orientation (its product is essentially in the consumer market) will be an important consideration in developing this plan.

First, Cirtem must make a marketing survey to determine which markets will support this new, advanced, metric product. The survey should be divided into two segments: the U.S. market and the foreign market. In each case, the first people to survey are Cirtem's salesmen and dealers, who deal with customer wants and needs daily. The reactions of salesmen and dealers will raise problems (servicing, for example) whose solutions should be included in the final marketing plan. Ultimately, a survey of consumer opinion will be necessary.

Selling metric conversion to the people who sell Cirtem products in the U.S. will be a consideration. Some dealers may be opposed to metrication just because it is something different. Cirtem must explain to these people what a metric product is and how they can put this to their advantage. For example, Cirtem might tell its salesmen and dealers that a metric product can more successfully compete with Cirtem's foreign competitors. By going metric, they are preparing for the future and avoiding product obsolescence.

Product design could provide a more significant advantage. If the tractor and accessories are conventionalized, Cirtem could include a metric tool with each product, which would ease the servicing problem. This is already being done with nonmetric products, and including metric tools with Cirtem's new product could compensate for any disadvantages metrication might cause. Supplying similar tools to Cirtem dealers would be another step in the right direction.

When, where, and how to announce the new product are equally important considerations. If Cirtem decides to produce two products (one inch and one metric), there will have to be two announcements, one to the U.S. market and one to foreign markets. In addition, two products will mean two operating manuals (and increased costs).

Lead time for training salesmen, technical service people, and dealers is another element in timing the release of Cirtem's metric product.

Even if it elects not to introduce a metric product in the U.S. at this time, Cirtem will have to do so sooner or later. At some point down the road, Cirtem's management will have to ask themselves: when will the U.S. be ready for a metric tractor? They might possibly decide that the U.S. consumer could care less whether Cirtem's tractor is inch or metric as long as it does the job and can be easily serviced. For this reason, Cirtem may or may not want to emphasize the new product as metric in its U.S. announcement.

Finally, Cirtem must face the question of product price: will a metric product mean an increase in the price to the consumer? If this is true, Cirtem's marketing plan should include provisions for dealing with the increased cost.

Decision-Maker's Metric Management Checklist

1. Marketing Metrication: The Survey
   a. Conduct market surveys for the U.S. (inch) and foreign (metric) markets.
   b. Survey salesmen, dealers, and technical service people for their reaction to metric products.
   c. Retail customer sampling is essential.
2. Marketing Metrication: The Advantages and Disadvantages
   a. In foreign markets, stress the advantages of metric products in company: announcements, advertising, publicity, dealer sales aids, consumer brochures and instructions, the dealers' service manual, and dealer training.
      • Emphasize the interchangeability of parts advantage to dealers and the servicing advantage to consumers.
   b. In the U.S. market, work to eliminate neutral or unfavorable attitudes toward metric products.
      • Conventionalize products and supply a standard tool(s) so that the consumer can make minor repairs himself. (This must be coordinated with Engineering.)

- Explain metric sizes carefully in the dealers' service manual.
- If the expense is justified, supply the required metric tools to dealers.

  c. In all markets, plan spare parts backup in the line of distribution so that dealers do not have to hold excessive inventory.

- This feature should be emphasized to dealers and consumers.

  d. Stress that metrication of the product has caused no increase in price.

- If it has, stress the overall cost advantages resulting from the interchangeability of parts.

3. Marketing Metrication: Timing

  a. Determine market (U.S. and foreign) support for metric products.

  b. Determine lead time for:

- Literature and advertising
- Training salesmen, technical service people, and dealers

  c. Equip the order processing system to handle customary and metric units.

- Both systems will continue to be used for a number of years.

Case Study Results 6

METRICATION OF COMPUTER SYSTEMS

Summary

A quick look at any metric practice guide will show that, contrary to what some say, conversion of Cirtem's computer systems can cause many problems. Those who believe otherwise need a basic education course in the SI metric system. The rules for spelling, abbreviating, and punctuating metric units and terms will all affect Cirtem's computer operations.

For example, input/output (I/O) devices handle only Roman capital letters. This will cause problems with the accepted symbols for Mega (M), milli (m), and meter (m); Giga (G) and gram (g); Kelvin (K) and kilo (k); etc. (Limited character set problems can be resolved, to some extent, by using ISO 2955-1974, Information Processing–Representation of SI and Other Units in Systems with Limited Character Sets.)

However, if Cirtem Systems people try to spell the metric units and terms out, they will only cause another problem by increasing the size of their field and the number of lines in their program.

The use of one master program to convert all of Cirtem's programs to metric will result in a loss of design integrity and invite errors in rounding. The following example illustrates the confusing consequences of the one program method. The Commonwealth of Pennsylvania used one program conversion for its computer operations with one result being that a four-footed animal ended up with 30 centimeters.

On the other hand, if Cirtem decides to convert in the core instead of with one program translation on I/O devices, there might be a problem finding core storage space for the additional information.

Other problems Cirtem will face are: computer calculations (mathematical, not analytical); storing, using, and comparing historical data with current (new) data; correcting rounding errors caused by conversion; modification of

166

computer design; and dual description of parts and products on bills of material, process routings, inventory listings, and other documents.

The first step in overcoming problems in converting Cirtem's computer operations is a basic education program in metric practice and the rules for writing in the SI metric system. ANMC's *Metric Editorial Guide* is a good starting point for this education program. It is especially important that Systems people receive this training, because learning the rules of metric practice will lead to an awareness of the problems that have to be solved.

There is also a need for an extensive systems analysis to determine the overall impact of conversion. This analysis should include the following considerations: overall time of the conversion program; coordination between Systems and the functional using areas; and cost estimates. The users rather than Systems should be given the responsibility for the needs survey because they will be more aware of the effect of metrication on day-to-day computer operations.

Decision-Maker's Metric Management Checklist

1. Computer Metrication: Education and Needs Analysis
    a. Educate Information Systems people in the rules of metric practice.
- Base this education on the metric system and metric practice guide approved by the company metrication committee.
- Start the education program early.

    b. Conduct a user survey of day-to-day computer operations affected by metrication and report the findings to Information Systems.
- Systems can then modify or redesign the programs as necessary.

2. Computer Metrication: Hardware Problems
    a. In general, basic hardware will be unaffected by metrication and will not need to be changed.
- However, I/O devices will be affected and may need to be modified.

    b. Consider alternative character sets available for I/O devices.
- I/O manufacturers can provide chains, trains, keyboards, fonts, balls, etc. which have the desired characters.
- Refer to ISO 2955-1974, Information Processing—Representation of SI and Other Units in Systems with Limited Character Sets.
- Alternate symbols, though undesirable in some cases, may be used.

3. Computer Metrication: Software Problems
    a. Both metric and customary units will be used for some time and will have to appear on the same documents in already existing fields and columns.

- Adopt a unit of measures code.

b. Carefully examine existing programs which handle both metric and customary units.

  - The rules of metric practice applied in old programs may be inconsistent with the metric practice guide adopted by the company metrication committee.

c. Add an identifying code so that the post processor can print the output in the desired units of measurement.

d. Programs involving built-in tables of customary values, sizes, constants, etc. will not be readily converted to exact metric values.

  - If accuracy is required, metric tables should be rewritten into the programs.

e. When using historical data for comparison purposes, convert customary units and terms to metric either manually or with a desk calculator.

  - Usually needs to be done on a one-time basis only.

METRICATION TRAINING

Summary

The difficulty and importance of selling metrication to top management has already been discussed. This must be accomplished before a metric training program can get underway. The training manager must know that there is a top-management commitment to metric. This means top management providing the funding, personnel, and other resources necessary to meet Cirtem's metric training needs.

One way of convincing top management is the hard-sell approach: a detailed training proposal written by Cirtem's training manager. This proposal would include Cirtem's training needs and objectives, as well as specific estimates of costs, time to be involved, etc. The report could compare the costs and impact of a well-planned training effort with a hit-or-miss approach to training.

However, as the case study illustrates, the hard-sell technique is not always effective. Since Cirtem's Vice President of Engineering, John Forbes, is still indifferent toward the need for metric training, a soft sell approach (sometimes called "managing your management") should be used.

For example, a self-test could be used to provide the impetus. Showing people what they know and what they do not know could motivate interest in metrication. However, the self-test method must be used wisely. Making it a metric trivia quiz could cause it to backfire. The self-test must be carefully developed and administered if it is going to work.

Resistance to a new system of measurement can be expected, but a successful metric awareness program will minimize this resistance. Cirtem employees must be informed of the decision to go to metric and convinced that the change will not be costly or disruptive—that it will be to their advantage. This is another sales job, and imagination is essential if it is going to succeed. Articles in the company newsletter, showing movies and slides during lunch, and metric posters are only three ways of developing metric awareness.

Cirtem is now ready to start the training job. A task analysis should be developed for each job category to determine which workers need what type and degree of training. Cirtem's training manager will realize that there will be different levels of training required, and he should develop a training program to meet these needs. Some type of modular method is useful.

Programmed instruction is the most effective method of industrial training, but it requires a substantial time investment in program development and writing. On-the-job training might prove more useful in certain job categories. The objective of any training method should be to provide Cirtem's employees with what they need to know, when they need to know it.

Cirtem's schedule for metric conversion will cause additional training problems. As more and more metric products are introduced, more and more Cirtem employees will have to be trained. For this reason, a canned training program is advantageous. Also, including a metric section in Cirtem's new employee training program will be helpful. Although this training might not be timely in terms of job application, it is an excellent opportunity to familiarize new employees with Cirtem's metric policy.

When metric legislation is enacted, Cirtem's training program will be supplemented by government and private education programs that will develop as a result of increased interest in metrication. However, Cirtem's metric coordinator must remember that although this "free" training will make Cirtem's training job easier, it will by no means do the job for them.

Finally, Cirtem's metric training program must be validated. There is only a passage of information, not training, if the training effort is not validated. Cirtem must know that its training is working, that it is meeting metric training needs in a timely, cost-effective manner.

Decision-Maker's Metric Management Checklist

1. Metrication Training: The First Steps
   a. Get top management commitment and financial support for metrication training.
   b. Once a conversion schedule has been approved, start a metric awareness program to reduce resistance to metrication.
      • Inform all employees of company metric plans.

    c. Define the learner population and identify training needs.
- Categorize all employees by tasks.
- Conduct a training needs analysis.
- Remember that some employees will need more training than others.

    d. Based on needs analysis, select proper training methods.

    e. Estimate time and funding necessary to complete training.

    f. Submit estimate to metric committee for approval.

2. Metrication Training: Program Development and Implementation

    a. Use training modules to avoid undertraining and overtraining.
- Employees take only the modules they need, as indicated by the needs analysis.

    b. Conduct training just prior to on-the-job need.

    c. Training should continue throughout company metrication.
- If metrication is by new products only, employees will need training as they are assigned to new product programs.
- Canned training programs can save you time and money.

    d. Validate your training program.

Case Study Results 8

METRICATION COST MANAGEMENT

Summary

As was explained earlier, "how much is it going to cost?" will be management's first question about metric conversion. However, if Cirtem's management looks only at metrication costs, they will get only half of the picture, for metrication has its benefits as well. Again, metrication is a business decision, and Cirtem will make that decision only when it makes economic sense to do so. Given the worldwide trend toward metrication, the question really is: Can Cirtem afford not to go metric? The main task before the corporate metric committee is to minimize the costs and maximize the benefits of the changeover.

Overmanaging metric costs is self-defeating. Close estimating and tracking of costs can add to the cost of metrication. Also, costs estimates are usually inaccurate: experience has shown that actual costs are normally lower than estimated costs. For these reasons, any attempt by Cirtem's accountants to develop a detailed estimate of metrication costs should be discouraged. The higher the understanding of metrication, the lower the costs of it.

A separate metrication budget should not be used, for in the past metric budgets have acted as magnets for all sorts of costs. As it has for other companies, following normal accounting procedures will probably work best for Cirtem. Including an ignorance factor for intangible costs might also prove useful.

Time is a critical determiner of costs. If Cirtem decided to go fully metric in six months, they would run into all kinds of extra costs. However, if the changeover is to be accomplished over a period of years, costs such as replacing machine tools can be buried in the normal capital replacement budget. When Cirtem's corporate metric committee considers the alternatives before them, one approach will be least costly while several others will be

172

totally unacceptable in terms of cost. Careful consideration of these alternatives is a key element in developing a minimum-cost strategy.

In conclusion, metric "doers" rather than metric "philosophers" are important in overcoming metrication cost problems. Once Cirtem has made a commitment to go metric, it should proceed with a can-do spirit, realizing that metrication costs are a part of doing business. With careful planning and management, metrication will reap more benefits than costs. Cirtem would not commit itself to the changeover if this were not true.

Decision-Maker's Metric Management Checklist

1. Managing Metrication Costs: A Strategy
   a. Develop an understanding of metrication at all levels of management.
      - This is a prerequisite for effective metrication cost management.
   b. Remember that proper timing and an orderly changeover are the best minimizers of metrication costs.
   c. Get middle management to accept metrication as a part of their job, as an added dimension to their area of responsibility.
   d. Some idea of metrication costs is necessary for policy commitment, the metric product launch, and for businesslike movement through the long metric transition.
      - Don't overmanage metrication costs.
      - Tracking metrication costs can be self-defeating and can add needless expense.
2. Managing Metrication Costs: A Plan
   a. Conduct a preliminary assessment of metric impact on all company operations.
      - Use a sampling technique.
      - Estimate unit cost effect for a cross section of items.
      - Extrapolate for total cost estimate.
   b. Using these estimates, cover metric costs in individual budgets for all company operations.
      - Do not establish a separate metrication budget.
      - Provide special budgetary supplements for extraordinary metric expenses.
   c. Accounting should be done under normal controls and procedures.
      - Use profit or cost centers as control points.
   d. Require local managers to achieve their metrication responsibilities within the budget.

e. The first metric product should not bear a disproportionate share of the ultimate total cost for metrication.

3. Managing Metrication Costs: Do's and Don't's

DO'S	DON'T'S
Get management commitment	"Bootleg" the metric changeover
Proceed by plan	Just "let it happen"
Use regular organization	Create a separate apparatus
Apply normal accounting practices	Require special budgeting
Coordinate with customers and suppliers	Set timing and pace unilaterally
Progress via new products and programs	Convert existing items
Identify benefits and pursue them vigorously	Assume that benefits will accrue without effort

Part IV

METRICATION COST MANAGEMENT:
WHAT OTHERS ARE DOING

Part IV offers a sampling of opinion and data on the subject of metrication cost management at the company level. The excerpted material is listed numerically, and is cross-referenced with the References at the end of Part IV. Although the sampling is by no means complete, the material does represent a consensus on the subject.

EXCERPTS ON METRICATION COST MANAGEMENT

Reference 1

"Much speculation has occurred in the U.S. regarding metrication costs. A wide divergence of opinion exists on the matter. Our opinions reside on the optimistic side, where we feel the costs will not be intolerable. In fact, with adequate planning, we believe costs can be readily absorbed within near normal operating costs. Obviously, some added costs are involved. Our estimate for a hypothetical company, or division of a company, is shown below. Note that the total costs extend over the ten-year period and are based on the percentages of gross sales for one year. . ."

Total Costs Over 10-Year Period
Cost Estimate — Division "X"
Based on $100 Million Gross Sales/Year

	Costs: 10-year total	Percent of annual gross sales
Design and documentation	($250 000)	0.25%
Modification or replacement of machines and equipment	($150 000)	0.15%
Maintaining dual supply of tools, parts, and materials	($250 000)	0.25%
Modification or replacement of testing equipment and gages	($210 000)	0.21%
Education and training	($160 000)	0.16%
Miscellaneous (marketing, sales, legal, tax)	($100 000)	0.10%
Total	$1 120 000	1.12%

Reference 2

"What part of the (metrication) cost is auditable: . . . why do you want to know how much it costs? . . . presumably because we want to manage resources intelligently. But a sound program of metric change calls for many activities that require a change in what people do without changing the number of people to do it, or even reducing measurably the amount of other useful activity they can perform. There are of course, some potential explicit costs in labor: preparation of educational materials, for example, or dual dimensioning of certain drawings for a period of time. We found (in the U.S. Metric Study) that companies which had already made a substantial conversion, either here or in foreign subsidiaries, assumed that additional headcount was not required and further that productivity was not appreciably reduced in the process . . . .

. . . let me turn now to the IBM strategy for metrication . . . . In the spring of 1971, management announced internally that SI was to be the preferred measurement system for selected new products and the processes supporting them. Retroactive design changes were not to be made unless economically justified. A careful, deliberate, company-wide program would be undertaken by planning metrication on a product-by-product basis; cost center control would be retained, no new management structures would be needed and management control of metrication costs would be judged by the same measurements of profitability and frugality as other elements of cost management . . . .

Non-product areas such as administration, forms, fixtures, furniture and buildings, were not to be impacted unless and until affected by a national change. IBM's objective is to gradually increase SI usage in our products until, by 1982, SI is our predominant measurement system, . . ."

Reference 3

"You may well ask – what is all this costing and how is it being financed? There are of course costs but the costs are far out-weighed by the benefits but, in fact it has not proved practicable to provide what might be called a "credible estimate" of costs. This is not only a United Kingdom view. No country has been able to make such an estimate. The actual costs depend on a wide range of events and decisions throughout the whole economy during the period of change; and in many cases it is not really possible to divorce the costs from those involved in the normal replacement of machinery and equipment and from those concerned with the whole process of innovation. . . ."

"Although it has not proved practicable to make a total assessment of metrication costs for the economy as a whole some estimates have been made in the United Kingdom by particular industries and firms for their own planning purposes. Each has been unique . . . ."

"For example, the British Steel Corporation put the estimated cost of its metrication in the region of 0.07 per cent of turnover a year for three years . . . .

The National Coal Board estimated that its principal costs spread over a number of years would not amount to more than 0.25 per cent of one year's turnover . . . ."

"Staveley Machine Tools Limited make it much less than 1 per cent on annual turnover spread over a period of several years . . . .

It would of course be unwise to generalize on the basis of individual estimates based on assumptions. We have no actual figures to set against them but all the indications are that estimates overstate the out-turn . . . ."

Reference 4

"Particularly significant manifestations of fear of the unknown are the fear of the consumer that wicked manufacturers and retailers will take advantage of his ignorance by robbing him, e.g., in the unit pricing of commodities; and the fear of manufacturers and retailers that if they change and their competitors do not they will be disadvantaged.

The Board is charged with keeping under review and reporting to the appropriate authorities any attempts to take *unfair advantage* of the public in the course of conversion. As already indicated, no single case has been uncovered in which, prima facie, action of the type contemplated was justified. This has been confirmed at three meetings of representatives of consumers' organizations the Board has convened.

We attribute this situation to the responsible attitude taken by manufacturers and retailers, and to the near certainty that any such attempt would quickly be detected and given wide publicity by the media . . . ."

Reference 5

"We are currently capable and willing to quote customers who submit their requirements in metric terminology and/or standards. There will be no added cost for this technical service. The added cost of special tooling and raw materials will be reflected in the quotation on the product itself. Upon special request we will furnish engineering specifications in metric equivalents on any of our existing products . . . ."

Reference 6

"The United States Steel Corporation has announced that it is now manufacturing steel products in standard metric sizes . . . . Notable is the fact that the mill prices for these USS standard metric sizes are the same as the mill prices for equivalent customary dimensions and specifications . . . ."

Reference 7

"*Question:* What are the   costs of an extended period of conversion in duplicate production, inventory and spare parts, increased handling and accounting costs?"

Answer: Unfortunately, insufficient data exist to provide specific details in answer to this question. However, a cost study undertaken in the preparation of answers to a U.S. Metric Study showed that an estimated in-house net added cost of metrication for our product over the optimum period as a percent of the total value of our 1969 sales would be projected to be approximately 2.2%. The various contributing areas can be broken down as follows:

— Personnel Education	5%
— Engineering and Research & Associated Documentation	16%
— Standards Association Activity	2%
— Manufacturing and Quality Control	56%
— Records and Accounting	14%
— Warehousing	1%
— Sales and Service	2%
— Other	4%

Extract from a company internal report

". . . a hidden part of the metric conversion expense for an established manufacturing operation is in the area of converting complex integrated management information, production scheduling, plant operation, accounting, inventory control and product quality reporting systems that support the management of current manufacturing facilities . . . ."

Reference 8

"A final critical view was that the transition would cost IBM millions. The contrary statement, of course, was that it would save us millions. I'm a skeptic on the subject of cost. I find costs very much influenced by management attitudes and where the management attitude is 'let's go' the cost of going metric is minimal. Where the management attitude is very negative, the costs tend to be very high. I really don't think that anyone knows the cost of our total plan. The challenge is to find the minimum cost path in each significant decision area . . . .

The name of the game is cost avoidance. We know that if we sit and just let the tide carry us out we'll pay a cost penalty. Again, if we go out too early we're going to pay a cost penalty for moving prematurely in some areas . . .
. . . our discipline in (finding the least cost path) is to charge each one of our divisions, within the corporate guidelines that have been established, with coming up with their specific plan to go metric. So, we are using the discipline of corporate/divisional planning to make sure what the best path is . . . ."

Reference 9

"Going metric efficiently is a function of good management. Bad management will postpone positive action and 'let it happen,' ultimately perpetuating the myth that the cost will be prohibitive . . . . . . one of the main lessons we have learned is that almost all costs were over-estimated, generally because they cannot be separated from the routine costs of carrying on a progressive business . . . individual companies do need to make an assessment; for example, the Dunlop Company expects the cost to be £3.50 per employee per annum over a planned eight years . . . ."

Reference 10

"A leading newspaper is to feature, during the next two years, a monthly review of prices of 230 consumer products, ranging from beer and petrol to goods on supermarket shelves. The newspaper sees the review as a means of deterring manufacturers and retailers from taking advantage of conversion to metric packaging . . . .

In the Age's first "Metric Watch" the Chairman of the Metric Conversion Board, Mr. J. D. Norgard, said: "We will follow up any complaints and, if necessary, refer them to authorities like the Consumer Protection Bureau . . . ."

Reference 11

". . . one large multi-billion-dollar electrical company, after careful study, reported a nominal 300 million dollars as its cost of converting to metric units; the cost would be higher if the program were confined to a ten-year period and lower if spread over 17 years. . . . Interestingly, another . . . corporation of approximately one-third the size reported metric conversion cost of $100 000 000." (p. 307)

Reference 12

". . . one must assess costs in relation to benefits. Does transition to the metric SI system and its costs result in the long run in benefits that outweigh the costs? The answer in our experience is that it will . . . . We can marshal many, many statistics on all aspects of the transition effort to show costs. However, our accumulated experience suggests that costs are all short-term ones and that the long-term benefits far outweigh them in gains. . . ."

Reference 13

"An immediate concern is the cost of metric conversion. There are costs, but experience has shown that realistic cost estimates are difficult to determine. At General Motors Corporation they are finding that their actual costs are far below original cost estimates. . . . policy within the Ford Motor Company on costs is just this: We told the divisions they were getting no budget relief for implementing the metric system. They operate on a Profit Center type operation, and we concluded it would cost us more money to try to track metric costs than it was worth . . . ."

Reference 14

". . . Ford . . . notes that the costs of such a (metric) change are really difficult to allocate . . . costs anticipated for a change to metrics are often grossly overestimated . . . ."

Reference 15

"I believe we have learned one significant thing from these studies, and that is that the cost of conversion should prove to be considerably less than originally expected . . . ."

Reference 16

The following information indicates the methods available, the cost of conversion elements, the installation cost, and the time to do so for the most common machine tools.

Estimates of Machine Tool Conversion for Metric Working[a]

Type of machine	Methods available	Cost per unit[b]	Installation cost per unit[b]	Time to convert per unit[b]
Lathes	Dual reading dials	$60–$125	$25–$60	½–1 day
	New feedscrews and nuts (metric)	$40–$100		1–1½ days
	Digital readout (customary/metric)	$1200–$2000 (small lathes)	$250–$500	2–4 days
		$2000–$3000 (large lathes) (price per axis)	$250–$500	

Estimates of Machine Tool Conversion for Metric Working[a]

Type of machine	Methods available	Cost per unit[b]	Installation cost per unit[b]	Time to convert per unit[b]
Milling machines	Dual reading dials	$90–$225	$30–$90	½–1 day
	New feedscrews and nuts (metric)	$90–$275	$60–$125	1–2 days
	Optical position indicators (metric)	$850–$1200 (per axis)	(included)	1–2 days
	Digital readout (customary/metric)	$2000–$3000 (per axis)	$125–$250	2–4 days
Horizontal borers	Digital readout (customary/metric)	up to $15 000 (3 axes)	$250–$325 (per axis)	2–5 days
	Optical position indicators	$500–$600	(included)	2 days
	New verniers	$250–$300	$25	½ day
Grinding machines	Engraving indicator drums	$20 per drum	$20 per drum	¼ day
Vertical borers	Dual dials	$200–$250	$215–$250	1–2 days
	Digital readout (customary/metric)	$1200–$2000	$60–$250	2–4 days
Drilling machines	Depth indicators (engraved)	$15 per plate	$15 per plate	¼ day
	Circular depth gage for radial-drill	$30	$15	¼ day
Planers Shapers Slotters	Dual reading dials	$75	$60–$125	1–2 days
Jig borers	Digital readout	$2000–$2750	$150–$300	2–4 days
	Optical position indicators	$1500–$2300	$300–$750	5–12 days
	Gage blocks	$1200–$1800	(included)	2–3 days

[a] This chart is taken from the June 27, 1975 *Metric Reporter.* It originally appeared in *Dun & Bradstreet's Guide to Metric Transition Management,* by Robert C. Sellers (Thomas Y. Crowell Company, New York, 1975, p. 113).
[b] Unit—means per unit, per screw, per axis, per machine, per drum, per plate as is appropriate for unit involved.

Reference 17

"Notes on budgets. It may be of assistance to those preparing a budget to read of the steps taken by one company* which has had experience in changing to metric working. In this case special attention was paid to discovering the most economic period in which to make the change.

The graph (A) shows the estimated expenditure calculated on various assumed times of change-over of the following budgeted items.

1. Purchase of new plant
2. Training
3. Stock holding
4. General overheads

These followed various patterns and are totalled and also recalculated using Discounted Cash Flow which takes account of the time money is tied up in stock not circulating, or the premature conversion of machines, or the training of staff ahead of the time when their knowledge is needed.

It also counts the following in addition to the purchase cost of items or work needed to change machines.

1. Cost of losses in cost recovery
2. Loss of profits due to loss of production
3. Loss of production when training workers
4. Salary of metric instructor

It should be noted that a crash program would inflate the cost inordinately. Also for this particular factory the optimum point for complete changeover is theoretically at about four years, but there is very little difference between three and five years.

It should be noted that the exact shape of the curves calculated in this *real* situation may be very different in other companies, but most of them can be expected to have these general characteristics. The cost of holding stock at the chosen time of 2½ years is shown as a separate graph, B . . . ."

"The calculations should be performed separately for various times say — 6 months, 1, 2, and 3 years. This is the only way to make a sound decision on the optimum duration of the change period. . . ."

* These graphs were kindly provided by Whiteley, Lang and Neill, Liverpool.

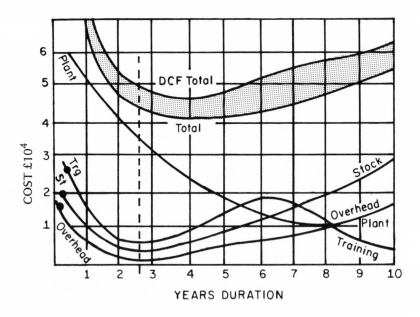

A.   Total cost for various durations of change.

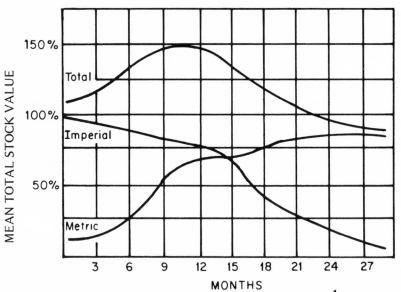

B.   Consumable stock value fluctuations for $2\frac{1}{2}$ yr program

Budget Studies of Cost of Change-Over to Metric Working in a Company Making Engineering Products

1. Description of Company—A high precision engineering company with 230 employees most of whom are machinists or toolmakers.
2. Product—Press tools, plastic moulding tools and special machines, jigs and fixtures.
3. Policy—It was decided to change over in 2½ years for the following reasons:
   a. All "teething" troubles would be over before 1975 and the company would be able to offer early metric service.
   b. Some of the Imperial equipment such as gauges and instruments which would have to be replaced in any case would have a market value if changed early. But it was accepted that there would be a need for and therefore an excess cost in changing a larger number of customer drawings from Imperial to metric as they would have to be made in this measure.
   c. Dual storing would cost less if a short period were adopted.
4. Cost of Change

	Cost	Percent
a. Conversion and testing machine tools	£ 40 000	58
b. Purchase of small tools and measuring equipment	11 500	17
c. Purchase of metric fasteners and taps	4 250	5.5
d. Purchase of drawing office equipment	50	1.0
e. Training and familiarization	4 250	5.5
f. Overheads & service (includes drawing conversion)	9 250	13
	£ 69 300	

These figures indicate that even in a case where the major part of the change-over cost is unavoidable, as it related to new machines, or alterations and the supply of new measuring equipment (in this case 75% of the total) training represents a fairly notable percentage of the cost and in absolute terms amounts to over £18 per head.

Reference 18

"Financial: A study was carried out to estimate the additional expenditure that metrication would necessitate. Expense estimates were grouped under five main headings:

(*i*)   cost of training,
(*ii*)   cost of additional scrap during familiarization period,
(*iii*)   cost of stocking additional components and raw materials,
(*iv*)   cost of small tools, and
(*v*)   cost of converting machine tools . . ."

Listed below are items of expenditure involved at the Grangewood experimental workshop, a typical experimental shop manned by four highly skilled fitter-machinists.

	£p
Change Gears: Smart & Brown Tool Room Lathe Model No. 1024	11·53
Collects: Smart & Brown Tool Room Lathes	27·66
Dials: Bridgeport Milling Machine. Adcock Shipley	60·37
Dial Test Indicators	16·00
Dies	7·75
Drills	16·62
Gauges: Feeler	0·67
Radius	1·76
Plug	16·23
Height	81·70
Helicoil Insertor Tools	15·41
Taps	9·17
Micrometers	79·90
Mandrels	12·23
Metric Blocks No. 3 Rubert 49 pieces	94·50
Parallels	33·00
Reamers: Hand	10·48
Machine	18·24
Rules & Scales	26·10
Spanners: Open End, Box, & Ring	19·26
Taps	44·67
Test Bars	61·61
Wrench Keys (to fit socket screws)	6·33
Total	£671·19

Reference 19

"*The relative priority of metric conversion* ... The normal processes of society—of industry, commerce, trade, education, government, agriculture, and all other activities—must go on while the language of measurement is progressively changed from customary to metric. The majority of those involved with conversion will plan and implement conversion in the context of their normal duties. Depending upon the organization, the goal may be a minimum-cost conversion or a maximum-benefit-for-change. In either case, the requisite is a nondisruptive transition, maintaining an effective operation while taking advantage of normal cycles of change . . ." (p. 126)

"*The costs of metric conversion* . . . . In summary, I really cannot tell you how to calculate the cost of going metric—in a company or in the U.S.A. as a whole. I can tell you that in most circumstances, the costs are sufficiently modest that there are other areas of cost control much more deserving of management attention. If conversion is planned well, management can pay for its cost many times over by paying attention to running the business better—rather than to running a metrication program from the corporate office. What is good advice for company managements probably is pretty good advice for the country too . . . ." (p. 151)

"There are companies for which the criteria for metric design of a product requires that there be *no identifiable cost increase* for the product. The design and production of the 2.3 litre Ford engine was carried out under these demanding requirements. General Motors projects a *zero-incremental cost* of new metric motor vehicle design and production. The no-metric-cost-increase condition will not be attainable except for new design; even under these conditions, there may be identifiable costs of conversion. The fact that a motor vehicle manufacturer finds it profitable to spend in excess of $100 million to set up a new line to produce a single engine to metric design demonstrates that it is meaningless to claim 'high costs of conversion' without stipulating the conditions under which the conversion is to take place . . . ."

"*The Australian experience* provides an example in which there were apparently no traceable price increases. Their Metric Conversion Board states that it is unaware of price increases being granted by the Prices Justification Tribunal on the basis of conversion costs . . . ." (p. 154)

Reference 20

"... the most thorough studies on the cost of metrication were conducted by companies engaged in the manufacture of transportation equipment, particularly automobiles and trucks. Reporting in industry 3711 are three companies, two of which are giants of the automotive industry of the world. We are well acquainted professionally with those responsible for submitting the reports and conducting the investigations ... we have a high degree of confidence in the estimates ... we deduce that the *cost of metrication* in the passenger car automobile industry over the period of transition would be about 6 percent of value added. If this cost is spread evenly over a period of 12 years ... and if the percentage added cost of metrication of suppliers to the automobile manufacturers is about the same as that of the manufacturers themselves, the cost of metrication borne by the consumer would be about *1/2 of one percent of sales value* ... it seems that these cost estimates, which are based upon very serious studies ... represent a realistic estimate of the cost involved in a fairly complicated product such as automobiles which account for a very large part of the gross national product ...." (p. xix)

"... cost estimates were plotted ... cost of metrication in terms of value added by manufacture is shown on a logarithmic scale because of the wide range in estimates. The abscissae ... are SIC industry numbers. (p. x)

Total Cost of Metrication as Percentage of Value Added by Manufacture in 1969

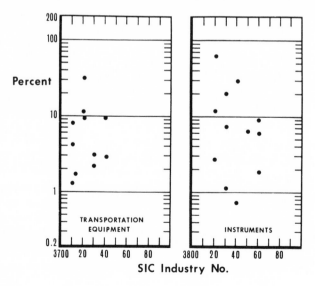

SIC Industry No.

"Transportation equipment manufacturers range over a factor of 23 in their estimates, the higher estimates coming from the aircraft industry. Estimates from the companies engaged in the manufacture of scientific and controlling instruments differ in their estimates by a factor of 87 . . . ."

Reference 21

"In March of 1974, McDonnell Douglas Corporation (MDC) undertook an *informal* study to determine the costs associated with converting to the use of the metric system. Corporate Management would simply not accept a "we don't know" answer to the question of "what are metrication costs?" Since requirements for conversion at MDC are expected to be on an individual program basis, two existing military aerospace programs were used as "study models"; the F-15 Air Superiority Fighter and the Harpoon Missile. Costs incurred through the design, development, and flight test or "pilot" production phases were estimated. . ." "Cost factors" assigning a unit figure to each metrication task were identified where possible. Following are examples.

Activity	Assumptions	Cost factor
Engineering		
Design	Engineering drawings will be prepared in metric dimensions	
	"Conventional" dual dimensioning will not be used	
	Standard "inch sized" drafting paper will be used initially	
Prepare conversion charts	Conversion charts showing inch/metric equivalent values for all lined dimensions on the drawing will be automatically computed and attached to each drawing for reference	$1-\frac{1}{2}$hr/dwg @ (A) = $X/dwg (1 dwg = 3ea 34″ x 40″ sheets)
	Eventually, conversion charts will not be needed	

Activity	Assumptions	Cost factor
Review and revise engineering handbooks	All manuals/handbooks will need review	
	Inch dimensions will be converted to metric units	
Design handbook		1 hr/sht @ $(A) = $X/sheet
Drafting room manual		1 hr/sht @ $(A) = $X/sheet
Material & process specification		3–½ engr.hr/spec @ $(A) = $X/sheet 2–½ cler.hr/spec @ $(A) = $X/spec Total $XX/spec
Scales and templates	Drafting machines will be equipped with metric scales	$X/drafting machine
	Design engineers will be provided with templates/scales	$X/Design Engineer $X/Supporting Engineer
Training		
Training classes	A minimum amount of formal training is required	2hr/eng @ $(A) = $X/eng
Standards	A full complement of hard metric standards suitable for aerospace use will not be available	
	Initially, majority of standards used will be inch standards	
	The mere availability of a hard metric standard will not be	

Activity	Assumptions	Cost factor
	sufficient reason to specify its use; performance, cost and availability will continue to be criteria	
	The translation of existing standards from inch to metric units is wasteful and will not be required	
Development of metric standards	Some new, "hard" metric standards will be required and and can be suitably developed, and others revised; $\frac{1}{3}$ of existing standards will be replaced by hard metric standards	
Fasteners		Engineer Effort 15hr/std @ $X(A) = $X/std
Mechanical Parts		Support Effort 4hr/std @ $X(A) = $X/std
Fluid Systems		
Electronic/Electrical		
Supporting Document		100hr/document @ $(A) = $X/ document

Note (A) = current applicable labor rate in $/hr including overhead.

Reference 22

Army Cost Estimates: "The projected cost of metrication of the Army is approximately $4.35 billion to be spent over a 30-year period. The highest costs are expected during the seventh through tenth years of the conversion. ... cost estimates are based on reports from 36 Headquarters organizations, extrapolated to represent the total Army budget. Most of these 36 reports were extensive compilations of subordinate activity and installation reports . . . ."

Shown below is a graph of the Army Budget Metrication Cost Profile

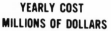

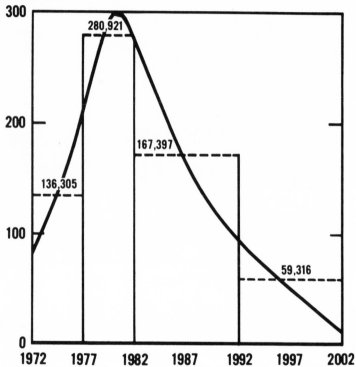

Reference 23

"The methods of conversion available are:
1. Replacement scales . . . Replace the present dial by a metric dial graduated in divisions representing convenient slide movements in millimetres. Alternatively, an additional metric scale can be provided on existing inch dials as shown below. If, for example, the feedscrew has 4 TPI, one revolution of the

handwheel and feed indicator represents 0.250 in or 6.35 mm. Alternatively, a 5 TPI feedscrew may have a dial with 200 graduations of 0.001 in. This corresponds to a feed per revolution of 5.08 mm, and it is possible to substitute or add a scale giving direct metric readings. In this case the back edge of the dial is graduated in increments of 0.02 mm and an extended zero indicator is fitted—the whole costing around £5 . . . ."

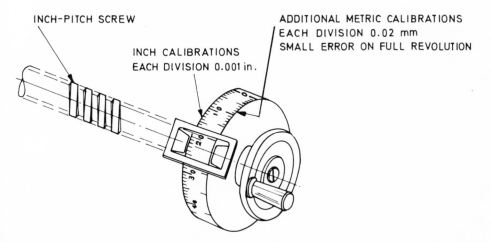

The Addition of a Metric Scale on a Existing dial Calibrated in Inches

Reproduced with the permission of the Controller of Her Britannic Majesty's Stationary Office, London, England.

Within a manufacturing company, one of the most controversial issues is whether or not inch conversions should be provided for metric dimensions and tolerances on engineering drawings.

Proponents say that, in order to successfully move metric designs into an essentially inch-oriented manufacturing and supply system and achieve timely, quality production on an economic basis—inch equivalents of dimensions must be available for reference as needed . . . and that, whether done by complete dual dimensioning or by application of a computer-generated chart, the conversions should be dropped, selectively, when no longer needed by those who use the drawings.

Opponents argue that—when metric transition is the object—dual dimensioning (whether in chart form or otherwise) is an "expensive exercise in futility"

... they say U.S. industry metric working capability already is widespread
and is growing rapidly—and that a company's production plant or a parts
supplier will quickly learn to use a "metric only" engineering drawing if that
is all they are given. In essence, the contention is that conversion charts are
not necessary—and, if provided, will prolong the changeover period.

Reference 24

"Drawing practices: This section recommends a uniform method of dimensioning and tolerancing drawings in SI measurement units. The report also
provides guidance for conversion of the drawing data between SI and customary units required during the transition period ... conventional dual
dimensioning of a drawing allows the part to be produced in either system of
measure, but experience has proved that it does not influence or encourage
an inch-oriented draftsman, engineer, or user of the drawing to achieve the
goal of adopting SI. Also, dual dimensioning adds clutter to the drawing and
would be difficult to remove when the transition period is over.

A metric dimensioned drawing without any conversion is the eventual goal,
but is very difficult to implement initially. Conversions of SI units will be
required during the transition period to eliminate repetitive conversions of
the drawing data by the many users. Listed below are a number of reasons
that justify the need for conversion data:

(a)   Engineers will require conversions to quickly check out interface
      dimensions of parts dimensioned in both SI and customary units.
(b)   Many existing data processing programs will not handle both measure-
      ment systems.
(c)   Suppliers of purchased finished parts or rough stock may not have
      made a decision to convert and may supply in customary units only.
(d)   Numerically controlled machine tools and some others, such as jig
      borers, that are calibrated in inches will require conversion data until
      the machine is converted or replaced.
(e)   Gages and tooling identification in either system of measure are easily
      recognized.
(f)   Conversion, made only once by people familiar with the precision
      intent on the drawing, will reduce the possibility of error.

Based on the preceding reasons, drawings should be dimensioned in SI with
conversions to customary units provided in chart form on the drawing. This
approach will permit easy removal at some point in the future and will help

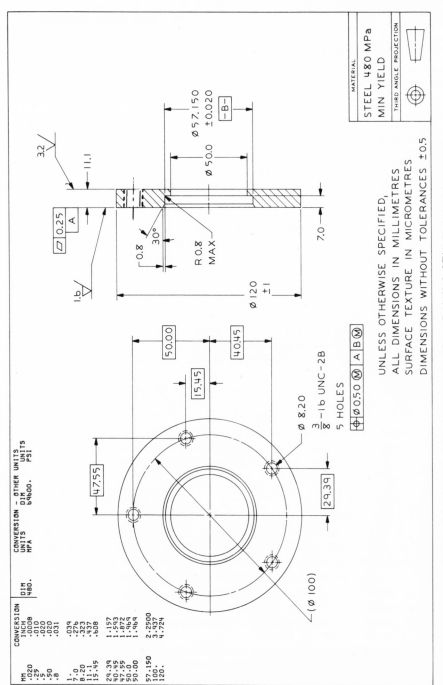

SAMPLE DRAWING WITH RECOMMENDED DRAWING STYLE AND PRACTICE

Reprinted with permission, Copyright © Society of Automotive Engineers, Inc., 1974, All rights reserved.

197

the user of the drawing to become familiar with and encourage his use of SI, since he must read the metric unit on the detail drawing before he can locate the conversion. . . ."

References

1. Foster, Lowell W. "The Honeywell Approach to Metrication," *Honeywell Computer Journal,* Honeywell Corporate Standardization Services.
2. Branscomb, Lewis. "Cost Implications of Increasing Metric Usage," in "Metrication–Cost Considerations," *Metric Conversion Paper #2,* American National Metric Council, August 1974.
3. Barnes, James. "The Financial Aspect of Metrication," in "Metrication–Cost Considerations," *Metric Conversion Paper #2,* American National Metric Council, August 1974.
4. Harper, A. F. A. "Metric Conversion in Australia: Planning and Public Acceptance," *The Canadian Business Review,* Spring 1975.
5. McCook, T. J. "Position on Metrication," (To: Alemite and Instrument Distribution Centers – U.S.A. and Canada), Stewart-Warner Alemite & Instrument Division, July 16, 1975.
6. "Metric Reporter," American National Metric Council, March 21, 1975, p. 1.
7. "Canadian Motor Vehicle Manufacturers' Association Response to Canadian Metric Commission," September 1973.
8. Rankine, L. John. "Company Problems in Converting to the Metric System," National Association of Manufacturers, October 12, 1972.
9. Lowe, R. W. "What Can Be Learned from the English Changeover," American Society of Mechanical Engineers, May 11, 1972.
10. "Newsletter," Australian Metric Conversion Board, January 1973.
11. "Should the United States Adopt the Metric System?" *Congressional Digest,* The Congressional Digest Corporation, December 1971.
12. Sellers, Robert C. (letter to Senator Warren G. Magnuson),"Metric Trauma? It's Not That Bad," *American Machinist,* January 7, 1974.
13. "American National Metric Council Statement to the House Subcommittee on Science, Research and Technology," *Transcript of Hearings on Metric Legislation.* May 6, 1975.
14. "Metrication: Industry Accepts the Challenge," *Automotive Industries,* July 1, 1973.
15. Foster, Lowell W. "Plan Now for a Painless Transition to Metric," *Assembly Engineering,* November 1972.
16. "Metric Reporter," American National Metric Council, June 27, 1975, p. 7.

17. Hellard/Connolly, Baden. *Metric Change,* London: Kogan Page Ltd., 1971.
18. "Metrication for Engineering Management," *A Metrication Board Report,* Urwick Technology Management Ltd., London: Her Majesty's Stationary Office, 1971.
19. Wertz, J. E. (principal investigator). "Metric Transition in the United States," *A Policy Assessment Study,* performed under Contract GI 40445 of the National Science Foundation, Rann Division, with the Regents of the University of Minnesota, March 1975.
20. "The Manufacturing Industry," *U.S. Metric Study Interim Report,* National Bureau of Standards Special Publication 345-4, July 1971.
21. Rau, R. W., and Neiner, J. J. "Metric Conversion Costs: A Study Conducted by McDonnell Douglas," *Metric Conversion Paper #6,* American National Metric Council, May 1975.
22. "Department of Defense," *U.S. Metric Study Interim Report,* National Bureau of Standards Special Publication 345-9, June 1971.
23. "Metrication in the Machine Shop," *A Metrication Board Report,* Machine Tool Industry Research Association, London: Her Majesty's Stationary Office, 1971.
24. "Recommended Guidelines for Company Metrication Programs in the Metalworking Industry—SAE J1066," *Handbook Supplement HS J1066,* Warrendale, Pa.: Society of Automotive Engineers, July 1974.
25. Benedict, John T. "A Pragmatic View of Metrication," a Panel Discussion, Warrendale, Pa.: Society of Automotive Engineers, February 26, 1975.
26. Standards Council of Canada. "Correct Usage of SI Units and Symbols—Common Errors," *Consensus,* July 1975.
27. American National Standards Institute. "Orientation for Company Metric Studies," March 1, 1970 (2nd edition).
28. Nance, C. T. "A Philosophy of Standardization for the 1970's," *British Standards Institution News,* January 1974.
29. Society of Automotive Engineers. "A Guide to Converting Drawings to Metric," *Automotive Engineering,* December 1973.
30. *Metric Conversion in Engineering and Manufacturing,* American National Metric Council, May 1974.
31. "Metrication—Legal and Trade Implications," *Metric Conversion Paper #1,* American National Metric Council, August 1974.
32. "Metrication—The Consumer Impact," *Metric Conversion Paper #3,* American National Metric Council, August 1974.
33. "Metrication—Problems and Opportunities," *Metric Conversion Paper #4,* American National Metric Council, August 1974.
34. Perica, Lou. "Impact of Metric Conversion on Technical Communication," *Metric Conversion Paper #5,* American National Metric Council, August 1974.

35. *A Report to the Nation on the Management of Metric Implementation,* American National Metric Council, March 1975.

36. Specht, Harry M. "Development of Metric Standards for Wire," *Metric Conversion Paper #7,* American National Metric Council, May 1975.

37. *Metric Editorial Guide,* American National Metric Council, August 1975 (2nd edition).

38. In-Service Training Sector Committee, *Metric Education Guide for Employee Training,* American National Metric Council, Ocotber 1975.

39. *Guidelines for Writers of SI Metric Standards and Other Documents,* American National Metric Council, January 1976.

40. Hastings, Russell. "ANMC Adopts Meter and Liter Spelling (Rationale and Survey Results)," *Metric Conversion Paper #8,* American National Metric Council, February 1976.

41. Prekel, Heinz L. "Metric Transition Management," *Metric Conversion Paper #9,* American National Metric Council, February 1976.

42. O'Brien, Richard J. "Metric Business Letter Sizes for North America," *Metric Conversion Paper #10,* American National Metric Council, February 1976.

43. *Metrication: Myths and Realities—Facing the Issues,* American National Metric Council, March 1976.

44. *Examining the Metric Issues,* American National Metric Council, May 1976.

45. "The Metric Conversion Act of 1975," *Metric Conversion Paper #11,* American National Metric Council, February 1976.

46. Heyd Kamp, Klaus. "Metrication of Computer Applications," *Metric Conversion Paper #12,* American National Metric Council, May 1976.

47. Ellard, Richard. "Metrication at Dunlap," *Metric Conversion Paper #13,* American National Metric Council, May 1976.

INDEX

American National Metric Council, 29, 54, 55, 58
  Metric Editorial Guide, 63n, 167
  Metric Education Guide, 86
  Metric Practice Committee, 63n
American National Standards Institute, 10, 58
  ANSI Z210.1, 38
American Society for Testing and Materials
  ASTM E-380, 36, 38
  codes, 36
American Society of Mechanical Engineers codes, 36

Cirtem Corporation, 103-107
Color code, 43, 161
Computer operations
  field requirements, 76, 78
  hardware, 70
  hardware conversion, 72-73, 167
  identifying code, 77
  ISO 2955-1974, 166, 167
  limited character sets, 72-73, 166
  mathematical calculations, 77, 168
  numerical control devices, 71-72, 76-77
  old metric usage, 76, 168
  rounding, 76, 78

software, 70-71
software conversion, 73-77, 167-168
systems analysis, 167
unit of measures code, 73-76
Conversion
  hard, 29
  of machines and equipment, 9-10, 45-46, 47, 159, 161
  soft, 29

Dual dimensioning, 37, 158, 195-198

Engineering
  design, 8-9, 39-40, 157, 158
  modular, 9, 40
  organization, 34
  product lifetime, 39-40, 157
  standards, 8-9, 36-37, 38-39, 158
Equipment replacement, 45, 47

Honeywell, Inc.
  ANSI Metric Advisory Committee, 14
  early metric interest, 11-16